COLT .45 REVOLVER AND SMITH & WESSON .45 REVOLVER M1917

FIELD MANUAL

BY WAR DEPARTMENT   October 20, 1941

DISCLAIMER:

This manual is sold for historic research purposes only, as an entertainment. It contains obsolete information and is not intended to be used as part of an actual operation or maintenance training program. No book can substitute for proper training by an authorized instructor.

FM 23-36

BASIC FIELD MANUAL

REVOLVER, COLT, CALIBER .45, M1917, AND REVOLVER, SMITH AND WESSON, CALIBER .45, M1917

Prepared under direction of the
Chief of Cavalry

UNITED STATES
GOVERNMENT PRINTING OFFICE
WASHINGTON : 1941

WAR DEPARTMENT,
Washington, October 20, 1941.

FM 23–36, Revolver, Colt, Caliber .45, M1917, and Revolver, Smith and Wesson, Caliber .45, M1917, is published for the information and guidance of all concerned:

[A. G. 062.11 (6–11–41).]

By order of the Secretary of War:

G. C. MARSHALL,
Chief of Staff.

Official:
E. S. ADAMS,
Major General,
The Adjutant General.

Distribution:
Bn and H (5) ; Bn 1 (10) ; C (10).
(For explanation of symbols see FM 21–6.)

TABLE OF CONTENTS

BASIC FIELD MANUAL

REVOLVER, COLT, CALIBER .45, M1917, AND RE-
VOLVER, SMITH AND WESSON, CALIBER .45, M1917

CHAPTER 1

MECHANICAL TRAINING

SECTION I

DESCRIPTION

■ 1. GENERAL.—*a.* The Colt revolver, caliber .45, M1917, and
the Smith and Wesson revolver, caliber .45, M1917, are single
shot, breech loading hand weapons. Each is provided with a
cylinder having six chambers arranged about a central axis
so that six shots may be fired before reloading is necessary.
Both weapons may be fired either single action or double

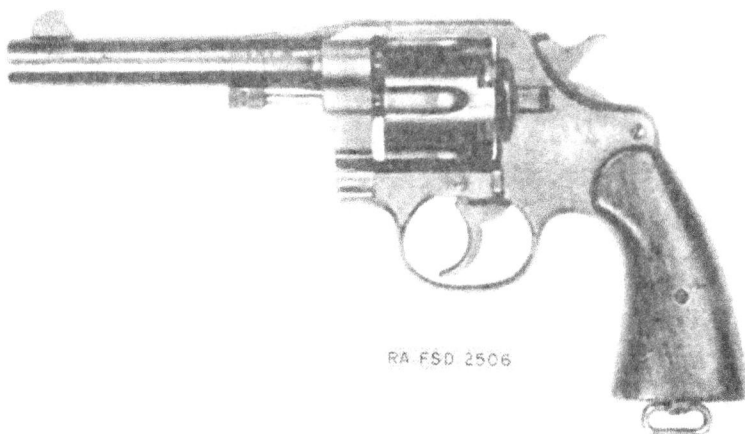

RA FSD 2506

FIGURE 1.—Colt revolver, caliber .45, M1917—assembled view.

action. Single and double action for these revolvers is described in detail in section IV.

b. These weapons are designed to fire the cartridge, ball, caliber .45, M1911. The revolver fires but once at each squeeze of the trigger. The action of cocking the hammer causes the cylinder to rotate and aline the next chamber with the barrel.

c. The rate of fire is limited by the dexterity of the operator in reloading the cylinder and the ability of the firer to aim and squeeze.

RA PD 2512

FIGURE 2.—Smith and Wesson revolver, caliber .45, M1917—assembled view.

■ 2. TYPES.—*a. Colt revolver, M1917.*

Weight_____pounds__ 2½
Total length _____inches__ 10. 8
Barrel:
    Length_____do____   5. 5
    Diameter of bore_____do____   . 445
    Diameter of rifling_____do____   . 452
Rifling, number of grooves_____ 6
Grooves:
    Width _____inches__   . 156
    Depth_____do____   . 0035
    Twist, one turn in_____do____ 16
Lands, width_____do____   . 073

Cylinder:
    Length _____inches__   1. 595
    Diameter_____do____   1. 695
Chambers:
    Number _____   6
Diameter:
    Maximum _____inches__   . 4795
    Minimum _____do____   . 473
Front sight above axis of bore_____do____   . 7325

b. Smith and Wesson revolver, M1917.

Weight _____pounds__   2¼
Total length_____inches__   10. 79
Barrel:
    Length _____do____   5. 5
    Diameter of bore_____do____   . 445
Rifling, number of grooves_____   6
Grooves:
    Width _____inches__   . 157
    Depth_____do____   . 003
    Twist, one turn in_____do____   14. 659
Lands, width_____do____   . 075
Cylinder:
    Length _____do____   1. 537
    Diameter_____do____   1. 708
Chambers:
    Number _____   6
Diameter:
    Maximum _____inches__   . 480
    Minimum_____do____   . 4795
Front sight above axis of bore_____do____   . 794

SECTION II

DISASSEMBLING AND ASSEMBLING

■ 3. DISASSEMBLING COLT REVOLVER (fig. 3).—*a.* (1) Remove crane lock screw (19) and crane lock (18) (fig. 3①).

(2) Press latch (20) (fig. 3②) to the rear, push cylinder to the left, and remove the cylinder and crane assembly by pushing to the front.

(3) Remove stock screw and stocks.

(4) Remove side plate screws.

(5) Remove side plate. Do not pry from its seating. With
wooden handle of a tool, tap the plate and frame until the
side plate loosens and lift out.

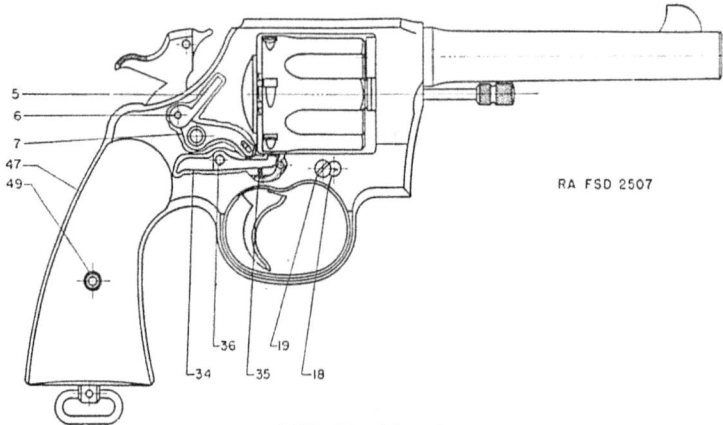

RA FSD 2507

① Right-side view.

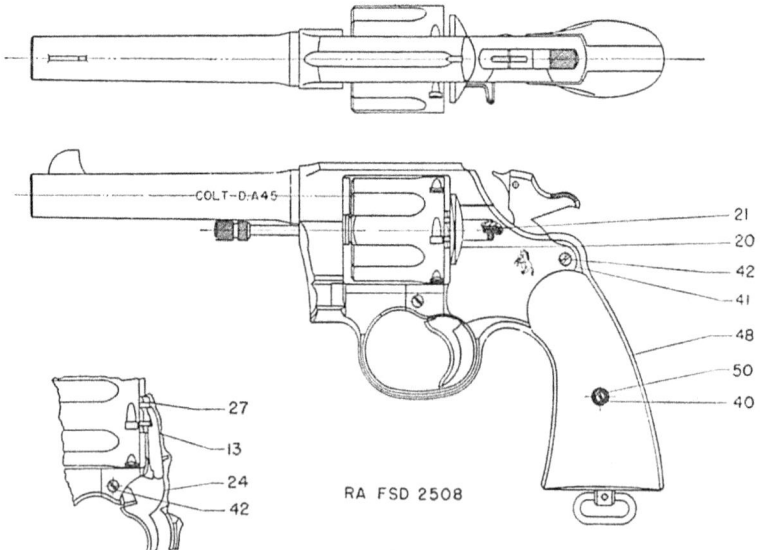

COLT-D-A45

RA FSD 2508

② Left-side view.

FIGURE 3.—Revolver, Colt, caliber .45, M1917.

4

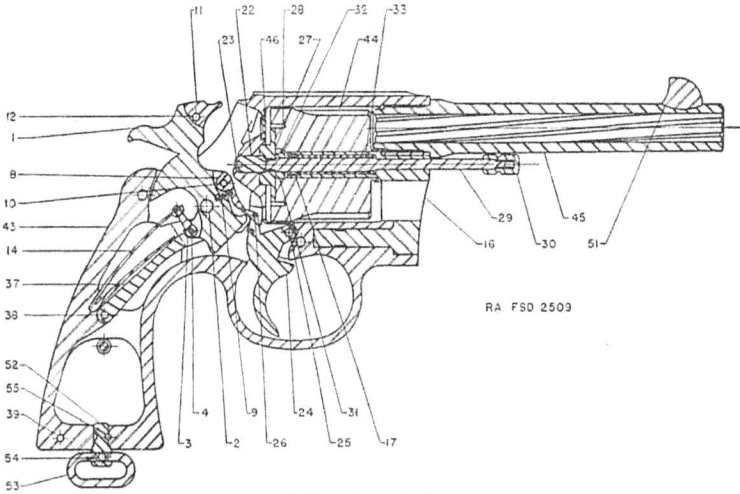

③ Sectional view.

FIGURE 3.—Revolver, Colt. caliber .45, M1917—Continued.

NOTE.—The latch and latch spring are removed with the side plate.

(6) Remove latch and spring from side plate.

(7) Remove mainspring by lifting the rear end from its seat, and disengaging the long end from the hammer stirrup (3) (fig. 3③).

(8) Remove the hand (13) (fig. 3②).

(9) With a drift, drive the rebound lever pin (38) to the right and remove rebound lever (37) (fig. 3③).

(10) Remove the trigger by lifting from the trigger pin (25) (fig. 3③).

(11) Draw the hammer to its rearmost position and lift from the hammer pin (2) (fig. 3③).

(12) With a small drift, drive out the strut pin (10) and remove the strut (8) and strut spring (9) (fig. 3③).

(13) Drive out the hammer stirrup pin (4) and remove the hammer stirrup (3) (fig. 3③).

(14) Remove safety lever (7) (fig. 3①) from its pivot.

(15) Remove safety (5) (fig. 3①) from its seat in the frame.

(16) Remove latch bolt (34) (fig. 3①) from its seat in the frame.

(17) Remove bolt screw (36) and lift out bolt (34) and bolt spring (35) (fig. 3①).

PIN—2

PIN—39

SCREW—40

RA PD 2510A

FRAME—43

PLATE—41

BLOCK—51

BARREL—45

RIVET—55

ESCUTCHEON—49

PIN—25

SCREW—42

ESCUTCHEON—50

STOCK—48

RIVET—54

RING—53

STUD—52

STOCK—47

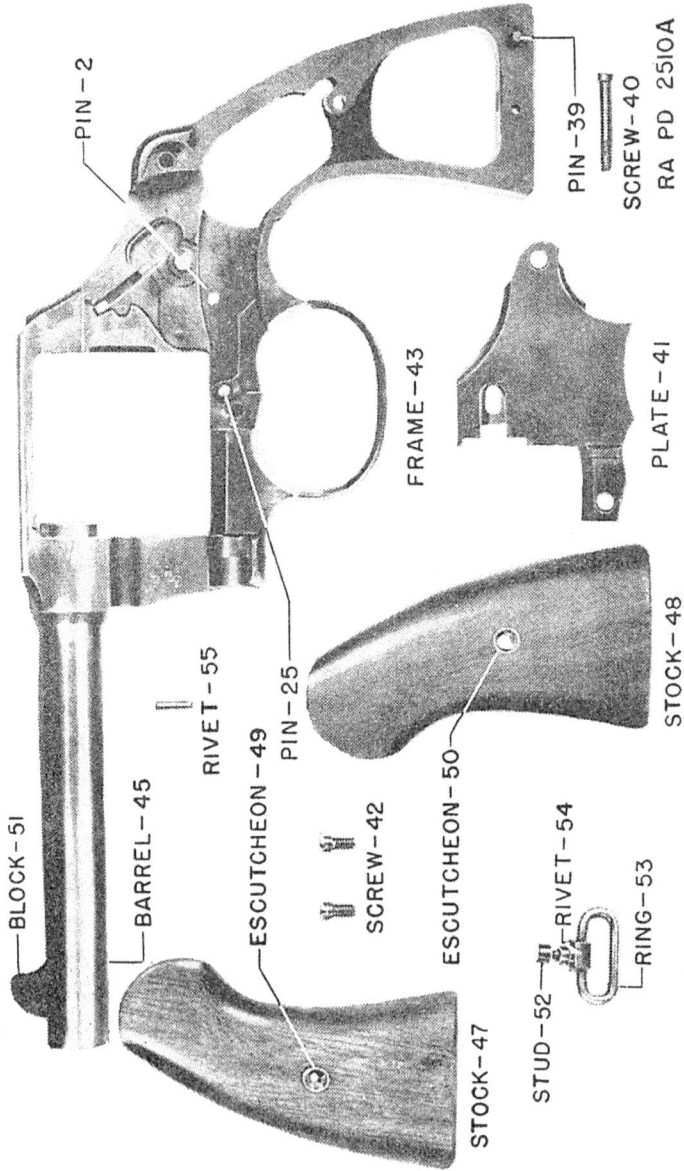

FIGURE 4.—Revolver, Colt, caliber .45, M1917, parts.

HEAD – 30

RATCHET – 27

SPRING – 33

PIN – 28

LATCH – 20

SPRING – 21

CYLINDER – 44

BUSHING – 17

RA PD 2511A

ROD – 29

PIN – 22

STUD – 23

CRANE – 16

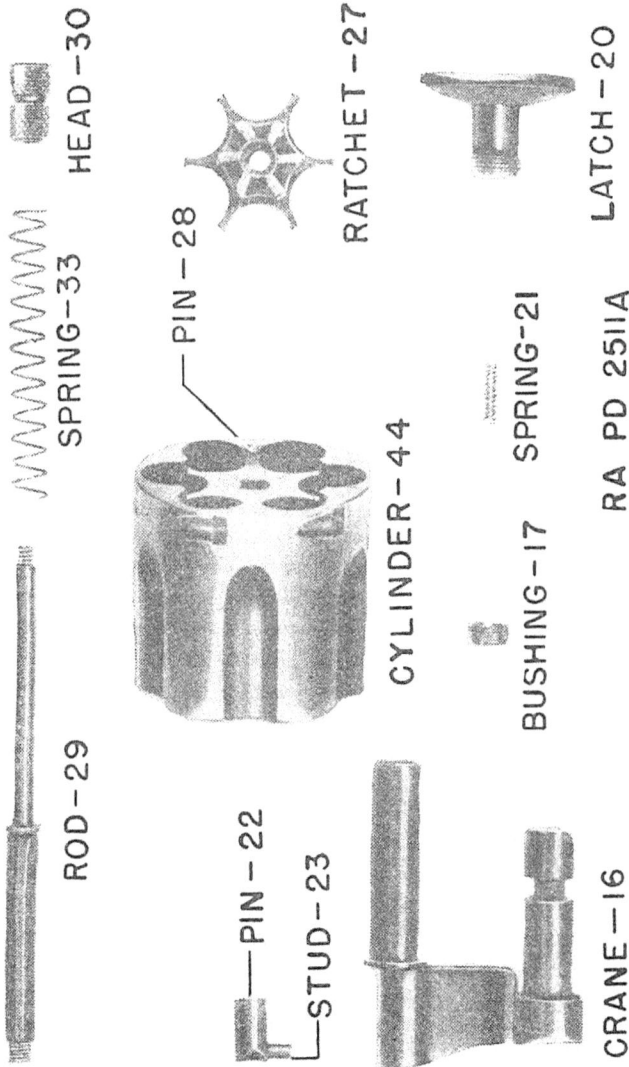

FIGURE 4.—Revolver, Colt, caliber .45, M1917, parts—Continued.

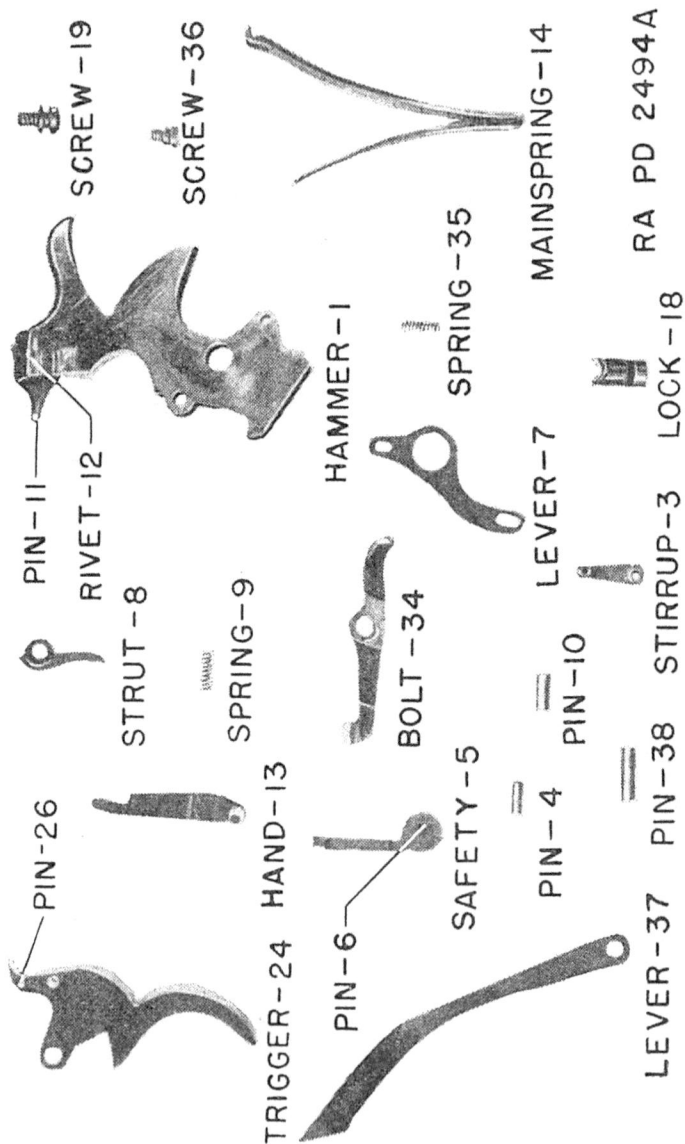

FIGURE 4.—Revolver, Colt, caliber .45, M1917, parts—Continued.

b. The following parts are disassembled by Ordnance Department personnel only:

(1) Barrel from frame.
(2) Trigger pin from frame (25) (fig. 3③).
(3) Stock pin from frame (39) (fig. 3③).
(4) Trigger and safety pin (26) (fig. 3③) from trigger.
(5) Firing pin (11) (fig. 3③) from hammer.
(6) Swivel ring (53) from swivel stud (52) (fig. 3③).
(7) Cylinder from crane (16) (fig. 3③).
(8) Ratchet (27) (fig. 3②) from ejector rod (29) (fig. 3③).
(9) Ejector spring (33) (fig. 3③) from cylinder.
(10) Head (30) (fig. 3③) from ejector rod.
(11) Escutcheons (49) (fig. 3①) and (50) (fig. 3②) from stocks.

■ 4. ASSEMBLING COLT REVOLVER.—*a.* Replace the cylinder bolt (34), cylinder bolt spring (35), and cylinder bolt screw (36) (fig. 3①).

b. Place the safety assembly (5) (fig. 3①) in its seat in the frame.

c. Place the safety lever (7) (fig. 3①) over its pivot with the slot in the short end engaging the stud on the safety.

d. Replace the latch pin assembly (22) (fig. 3③) in its seat in the frame.

e. Replace the trigger on the trigger pin so that the stud on the right side of the trigger engages the slot in the longer end of the safety lever (7) (fig. 3①).

NOTE.—Test by working trigger forward and back. If the safety lever and safety operate, the assembly is correct.

f. Assemble the strut (8), strut spring (9), and strut pin (10) (fig. 3③) to the hammer.

g. Assemble the hammer stirrup (3) and hammer stirrup pin (4) (fig. 3③) to the hammer.

h. Place hammer assembly in place on the hammer pin.

i. Assemble the rebound lever (37) to the frame with the rebound lever pin (38) (fig. 3③).

j. Replace the mainspring so that the notched end engages the hammer stirrup (4) (fig. 3③).

k. Insert the stud on the hand (13) (fig. 3②) in its hole in the trigger. Press upward on the rebound lever to permit the hand to be fully seated.

l. Replace the latch pin spring (21) (fig. 3②) in its seat in the side plate.

m. Put side plate in position but not fully seated.

n. Place the latch in its slot in the side plate so that the latch pin stud (23) (fig. 3③) engages in the hole in the latch.

o. Seat the side plate fully and replace the side plate screws.

p. Replace the cylinder and crane, crane lock, and crane lock screw.

q. Replace the stocks and stock screw.

■ 5. DISASSEMBLING SMITH AND WESSON REVOLVER (figs. 5 and 6).—*a*. (1) Remove stock screw and stocks.

(2) Remove the side plate screw near the forward part of the trigger guard.

(3) Press forward on the latch to release the cylinder. Push the cylinder to the left and withdraw cylinder and crane assembly to the front, being careful to prevent the crane stop pin (8) and crane stop spring (9) from flying out.

(4) Remove crane stop plunger and spring.

(5) Unscrew thumb piece nut (58) and remove thumb piece (57).

(6) Remove the remaining three side plate screws.

(7) Remove side plate. Do not pry side plate from its seating. With wooden handle of a tool, tap the plate and frame until the side plate loosens, and lift from its seating.

(8) Remove strain screw (56) from recess in butt end of frame.

(9) Remove mainspring (44) by pushing bottom end to the right from its recess in the frame.

(10) Remove rebound slide (45) and rebound slide spring (47) by pressing the rear end of the slide to the right until it clears the rebound slide pin (46).

NOTE.—Hold thumb over rear end of slide as it is removed from the pin in order not to lose the spring.

(11) Remove the hand assembly (33).

(12) Pull the latch (38) back until it clears the rear of the hammer and pull the hammer to the rear. It may be necessary to press the latch away from the frame to allow the hammer to pass. Lift the hammer off the hammer pin (27).

(13) Press trigger assembly to the right and remove from trigger pin.

(14) Remove cylinder bolt plunger screw (15) and cylinder bolt plunger spring (14) and cylinder bolt plunger (13).

(15) Lift cylinder bolt (11) from its pin and remove.

(16) Push latch to rearmost position and remove by pushing the rear end to the right.

(17) Withdraw latch plunger (39) and spring.

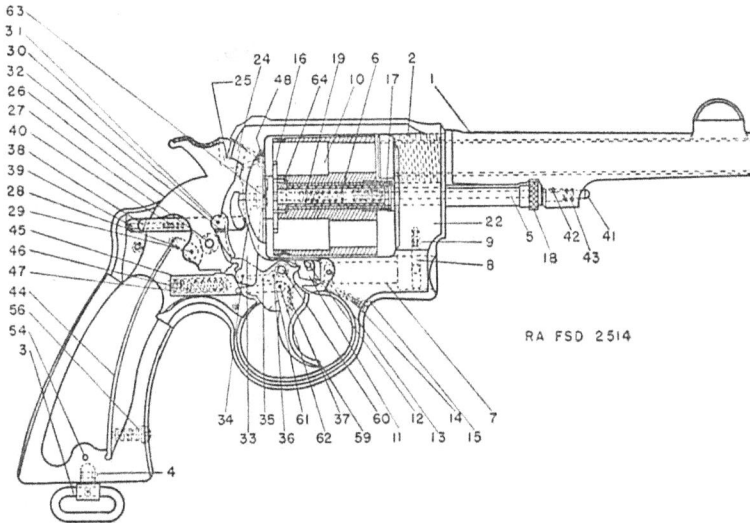

FIGURE 5.—Smith and Wesson revolver, caliber .45, M1917—sectional view.

b. The following parts are disassembled for repair purposes only by ordnance personnel:

(1) Barrel from frame.

(2) Stock pin (54) from frame.

(3) Rebound slide pin (46) from frame.

(4) Trigger pin (60) from frame.

(5) Cylinder bolt pin (12) from frame.

(6) Firing pin (24) from hammer.

(7) Escutcheons (20 and 21) from stocks.

■ 6. ASSEMBLING SMITH AND WESSON REVOLVER, M1917 (figs. 5 and 6).—a. Replace locking bolt spring (43) and bolt (41) with flat surface up. Replace locking bolt pin (42).

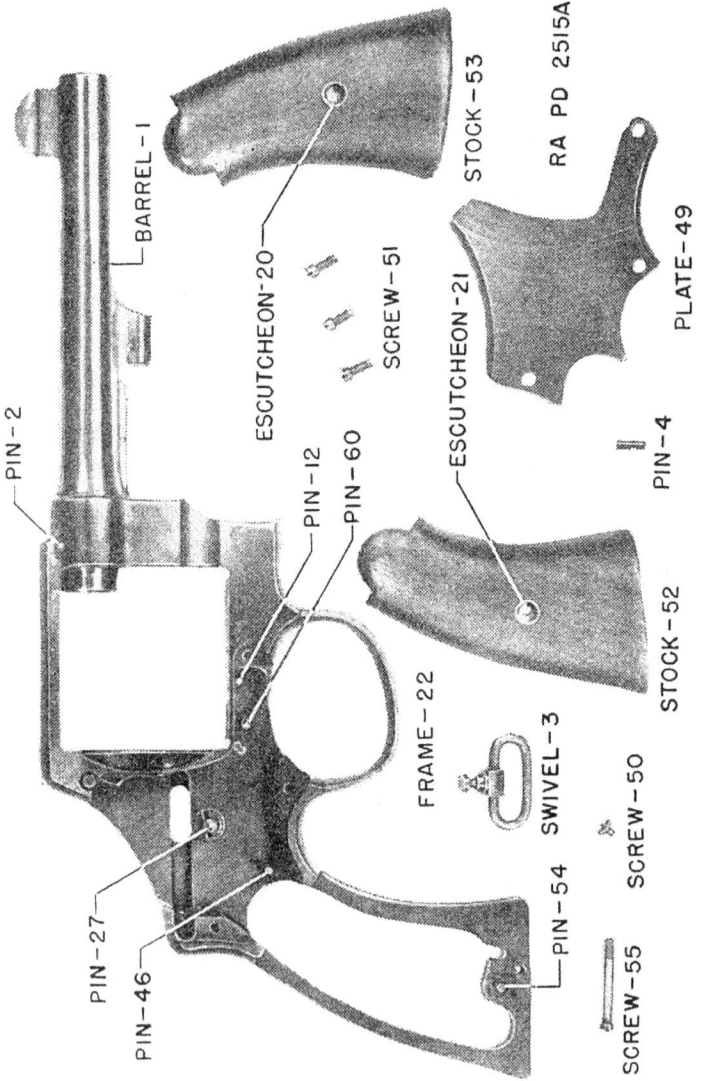

FIGURE 6.—Revolver, Smith and Wesson, M1917, parts.

ROD-5

PIN-64

BUSHING-63

EJECTOR-16

SPRING-19

RA PD 2495A

CYLINDER-10

COLLAR-17

SPRING-6

PLUNGER-18

CRANE-7

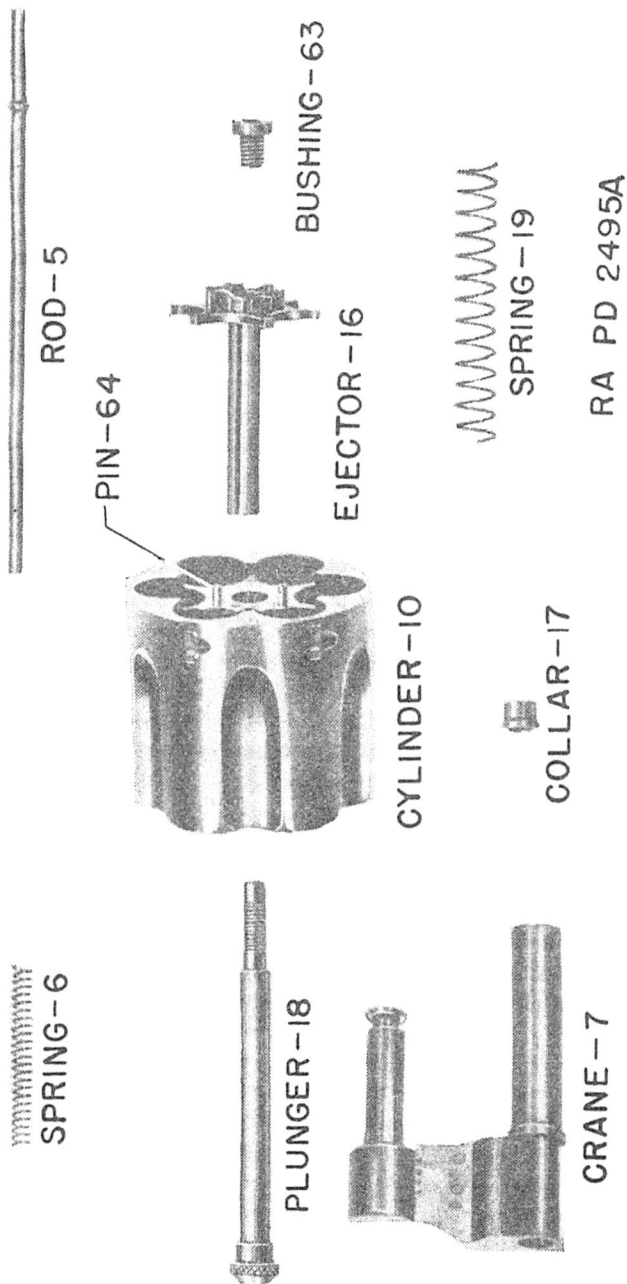

FIGURE 6.—Revolver, Smith and Wesson, M1917, parts—Continued.

13

SCREW—15

SPRING—37

SPRING—14

SPRING—40

LEVER—35

LEVER—61

SPRING—32

TRIGGER—59

PIN—8

STIRRUP—28   RA   PD   2516A

NUT—58

PIN—29

PIN—42

SCREW—56

BOLT—41

MAINSPRING—44

PLUNGER—39

PIN—36

STRUT—30

PIN—62

PIN—31   SPRING—9   BOLT—11

PIN—34

MAINSPRING—44

PLUNGER—39

SPRING—43   STIRRUP—28

PIN—24
RIVET—25

HAND—33

HAMMER—26

SLIDE—45

PIECE—57   SPRING—43

PLUNGER—13

SPRING—47

LATCH—38

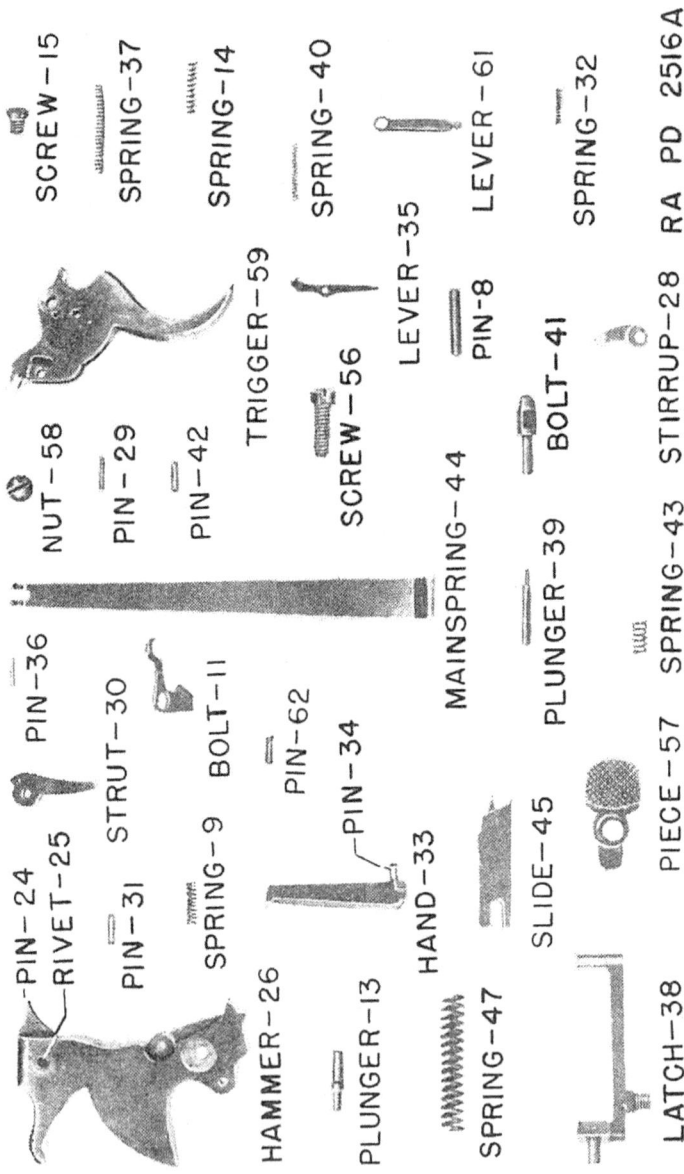

FIGURE 6.—Revolver, Smith and Wesson, M1917, parts—Continued.

14

b. Replace cylinder bolt (11) on its pin. Replace cylinder bolt plunger (13), cylinder bolt spring (14), and cylinder bolt screw (15).

c. Assemble the hand (33) to the trigger as follows: With the blade of a screw driver or drift, depress the forward end of the hand lever (35) against the hand lever spring (37). Place the hand pin (34) in its hole in the trigger so that the lug alongside the hand pin is engaged *below* the rear end of the hand lever.

d. Replace assembled trigger and hand on the trigger pin, holding the upper end of the hand to the rear to clear the frame, and with the rear end of the trigger lever (61) in its topmost position.

e. Replace the latch plunger (39) and latch plunger spring (40) in the recess in the rear end of the latch (38).

f. Replace latch in its guide in the frame by pressing the plunger (39) forward.

g. Replace the hammer assembly on the hammer pin.

NOTE.—To accomplish this the trigger should be in the rearmost position and the latch should be held to the rear.

h. Put rebound spring (47) into the rebound slide (45) and replace the assembly on the rebound slide pin (46) with beveled end forward, so that the rear end of the trigger lever engages the notch in the forward face of the rebound slide.

i. Replace the mainspring by engaging the hooks on the upper end with the hammer stirrup (28) and then pressing the lower end into its recess in the frame.

j. Replace the mainspring strain screw (56).

k. Replace the side plate and all side plate screws but the forward one.

l. Replace crane stop plunger and crane stop plunger spring (9) in hole in crane.

m. Assemble crane (7) and cylinder assembly to frame.

n. Replace the remaining side plate screw.

o. Replace the thumb piece (57) and thumb piece nut (58).

p. Replace the stocks and stock screw.

Section III

CARE AND CLEANING

■ 7. GENERAL.—*a.* Careful and conscientious work is required to keep revolvers in a condition that will insure perfect functioning of the mechanism and continued accuracy of the barrel. It is essential that exposed parts of the mechanism be kept cleaned and oiled.

b. The mechanism also requires care to prevent rust or an accumulation of sand or dirt in the interior. Revolvers are not usually disassembled for cleaning under ordinary conditions. After immersion in water, after contamination by gas, or if excessive amounts of dirt or sand get into the interior, the side plate should be removed and the mechanism cleaned, dried, and oiled. The side plate is removed only under the supervision of an officer or noncommissioned officer.

■ 8. PROCEDURE.—a. Care and cleaning of the revolver include the ordinary care to preserve its condition and appearance in garrisons, posts, and camps, and in campaign.

b. Damp air and sweaty hands are great promoters of rust. The revolver should be cleaned and protected after every drill or handling. Special precautions are necessary when the revolver has been used on rainy days and after tours of guard duty.

c. To clean the revolver, rub it with a rag which has been lightly oiled and then clean with a perfectly dry rag. Swab the bore and chambers with an oily flannel patch and then with a perfectly dry one. Dust out all crevices with a small, clean brush.

d. Immediately after cleaning, to protect the revolver, swab the bore and chambers thoroughly with a flannel patch saturated with oil, sperm, if available, or oil, engine, SAE 10 or SAE 30. Wipe over all metal parts with an oily rag, applying a few drops of oil to all exposed working surfaces of the mechanism. Oil applied in the openings for the trigger, hammer, latch, and cylinder lock will work into the mechanism.

e. After cleaning and protecting the revolver, place it in the revolver rack without any covering whatever. The use of canvas or similar covers is prohibited as they collect moisture and rust the metal parts. While barracks are being swept, revolver racks will be covered with a piece of canvas to protect the revolvers from dust.

■ 9. CARE AFTER FIRING.—a. When a revolver has been fired, the bore and chambers will be cleaned thoroughly not later than the evening of the day on which it is fired. Thereafter it will be cleaned and oiled each day for at least the next three succeeding days.

b. To clean the bore after firing, first open the cylinder and hold the revolver with the muzzle pointed downward, toward the operator, and hold the cylinder in its full open position. The cleaning rod with a patch soaked in hot water and issue soap, hot water alone, or cold water is inserted in the muzzle and moved forward and back for about a minute. During this operation the patch is frequently dipped in water to keep it wet. When the bore is wet, a brass or bronze wire brush should be run all the way through the bore, then all the way back three of four times. A water-soaked patch should again be pushed forward and back through the bore. Then wipe the cleaning rod dry, and using dry, clean flannel patches, thoroughly swab the bore until it is clean. Examine the bore carefully for metal fouling.

Caution: After firing do not oil the bore before cleaning. This will eliminate the difficulty of removing an oil deposit in addition to the removal of the powdered salt.

Note.—If available, authorized bore cleaning fluid should be used in lieu of the water solutions referred to above.

c. Repeat the operation in *b* above for each of the chambers of the cylinder, holding the revolver with the muzzle toward the operator, cylinder beneath the frame.

Caution: After firing do not oil the chambers before cleaning.

d. Saturate a clean flannel patch with sperm oil and swab the bore and chambers with the patch, making certain that the bore and all exposed metal parts of the revolver are covered with a thin coat of oil.

e. Due to corrosion caused from the gas which escapes between the barrel and the cylinder, the following parts require special care after firing:

(1) The frame just above the cylinder in the rear of the barrel.

(2) The nose of the hammer.

(3) The firing pin channel and the hammer groove in the frame.

■ 10. Rules for Care of Revolver on the Range.—*a.* Always clean at the end of each day's shooting. A revolver that has been fired should not be left overnight without cleaning.

b. Never fire a revolver with any dust, dirt, mud, or snow in the bore.

c. Before loading the revolver make sure that no patch, rag, or other object has been left in the barrel.

d. During range firing, a noncommissioned officer will be placed in charge of the cleaning of revolvers in the cleaning racks.

■ 11. CARE DURING EXTREME COLD WEATHER.—Use oil, lubricating, for aircraft instruments and machine guns, U. S. A. Specification No. 2–27, sparingly on the working parts after carefully removing all oil by washing with solvent, dry cleaning.

■ 12. CARE AFTER CHEMICAL ATTACK.—*a.* Revolvers should be disassembled and cleaned as soon as possible after a chemical attack.

b. Oil will prevent corrosion for about 12 hours.

c. Clean all parts in boiling water containing a little soda ash, if available.

d. All traces of chemical must be removed from ammunition with a slightly oiled rag; then thoroughly dry the ammunition.

e. Rust-preventive compound resists chemical corrosion more than light oil. In many exposures, especially those of long duration, ammunition treated with sperm oil evidences more severe corrosion than unprotected cartridges.

■ 13. IMPORTANT POINTS TO BE OBSERVED.—*a.* After firing the revolver, never leave it uncleaned overnight. The damage done is then irreparable.

b. Keep the revolver clean and lightly lubricated but do not let it become gummy with oil.

c. Do not place the revolver on the ground where sand or dirt may enter the bore or mechanism.

d. Do not plug the muzzle of the revolver with a patch or plug. One may forget to remove it before firing, in which case the discharge may bulge or burst the barrel at the muzzle.

e. A revolver kept in a leather holster may rust due to moisture absorbed by the leather from the atmosphere, even though the holster may appear to be perfectly dry. If the holster is wet and the revolver must be carried therein, cover the revolver with a thick coat of oil.

f. The hammer should not be snapped when the revolver is partially disassembled.

g. Pressure on the trigger must be released sufficiently after each shot to permit the trigger to reengage the hammer strut in single action firing, and to permit the trigger to engage the hammer strut in double action firing.

h. The side plate should not be removed except under the supervision of an officer or noncommissioned officer.

i. Never attempt to remove the side plate by prying it out of place.

j. The crane and cylinder of the Colt revolver must not be dismounted except by ordnance personnel.

k. Never attempt to open the cylinder when the hammer is cocked or partly cocked.

l. Never attempt to cock the hammer until the cylinder is fully closed and locked in the frame.

SECTION IV

FUNCTIONING

■ 14. METHODS OF OPERATION.—*a.* The chambers of the cylinder are loaded with six cartridges, either singly or in clips of three rounds. When the cylinder is closed the revolver is ready for firing.

b. In firing double action, pressure is applied to the trigger until the hammer falls, firing the cartridge.

c. In firing single action, the hammer is cocked by pressure to the rear with the trigger fully released. Pressure on the trigger releases the hammer which falls, firing the cartridge.

d. To lower cocked hammer on a loaded chamber without firing, draw hammer slightly to rear with the thumb; press the trigger to disengage from hammer; let hammer down slowly a short distance, and release trigger. Lower hammer as far as it will go.

■ 15. SAFETY DEVICES.—*a. General.*—It is impossible for the firing pin to discharge or even touch the primer except on receiving the full blow of the hammer.

b. Automatic safety devices.—(1) *Colt revolver.*—The safety lever which is pinned to the trigger moves the safety upward in front of the hammer when the trigger is released after firing a shot. The safety prevents the hammer moving forward sufficiently to strike the primer until pressure is

19

again applied to the trigger, thereby moving the safety downward out of the way. Thus an accidental blow on the hammer cannot cause the revolver to fire. The nose of the cylinder bolt actuated by the cylinder bolt spring projects through a slot in the frame and engages one of the rectangular cuts in the cylinder. This insures positive alinement of one of the chambers of the cylinder with the barrel.

(2) *Smith and Wesson revolver.*—A projection on the lower end of the hammer resting against the upper surface of the rebound slide prevents the hammer moving sufficiently far forward to strike the primer, except when the trigger is all the way to the rear. Thus an accidental blow on the hammer cannot cause the revolver to fire. The lug on the upper rear end of the cylinder bolt actuated by cylinder bolt spring projects through a slot in the frame and engages one of the rectangular cuts in the cylinder. This insures positive alinement of one of the chambers of the cylinder with the barrel.

■ 16. DETAILED FUNCTIONING OF COLT REVOLVER, M1917 (fig. 3).—*a.* The lock mechanism is contained in the frame and consists of the hammer with its stirrup, stirrup pin, strut, strut pin, and strut spring; the trigger with its pin; the rebound lever; the hand; the cylinder bolt with its spring; the mainspring, which also serves as a rebound lever spring, the hand spring, the trigger spring; the safety and safety lever.

b. The hammer and trigger are pivoted on their respective pins, which are fastened in the right side of the frame. The rebound lever is pivoted on its pin within the grip of the frame. The lower end of the mainspring fits into a slot in the frame, and its upper end engages the hammer stirrup.

c. The lower arm of the mainspring bears on the upper surface of the rebound lever, so that the latter, when the trigger is released after firing a shot, carries the hammer back to its safety position and forces the trigger forward, bringing the hand back to its forward and lowest position. The safety lever, being pinned to the trigger, moves the safety upward in front of the hammer by this same motion.

d. The revolver may be used either single action or double action. In firing double action, pressure upon the trigger causes its upper edge to engage the hammer strut and thereby raises the hammer until nearly in full-cock position, when the strut will escape from the trigger, and the hammer, under action of the mainspring, will fall and strike the cartridge. In firing single action, the hammer is first pulled back with the thumb until the upper edge of the trigger engages in the full-cock notch in the front end of the lower part of the hammer. Pressure on the trigger will release the hammer, which, under the action of the mainspring, will fall and strike the cartridge.

e. The bolt is pivoted on its screw, which is supported in the right side of the frame. The bolt spring pressing upward causes the nose of the bolt to project through a slot in the frame ready to enter one of the rectangular cuts in the surface of the cylinder. During the first part of the movement of the trigger in cocking the revolver, the nose of the bolt is withdrawn from the cylinder by the rear end of the bolt coming into contact with the lug on the rebound lever, permitting the rotation of the cylinder. The object of the bolt is to hold the firing chamber in line with the barrel, and also to prevent the cylinder making more than one-sixth of a revolution at the time of cocking.

f. The hand is attached by its pivot to the trigger, and as the latter swings on its pin when the hammer is being cocked, the hand is raised, revolves the cylinder, and serves with the bolt to lock the cylinder in proper position at time of firing, that is, the axis of the chamber containing the cartridge to be fired coincides with the axis of the bore of the barrel. The pressure of the rebound lever on the lug on the hand insures the engagement of the hand with the ratchet.

g. The cylinder has six chambers. It rotates upon and is supported on the central arbor of the crane. The crane fits into a recess in the frame below the barrel, and turns on its pivot arm, which rotates in a hole in that part of the frame below the opening for the cylinder, and is secured by the crane lock and crane lock screw. The ejector rod passes through the center of the arbor of the crane supporting the cylinder, and, projecting under the barrel, terminates in the ejector rod head. The ratchet is screwed on the rear end of the ejector rod with a right-hand thread and then firmly secured by upsetting the end of the rod. The ejector spring is coiled around the ejector rod within the cylinder arbor of the crane, the front end bearing on a shoulder of the rod

21

and the rear end on the crane bushing, which is screwed with a right-hand thread into and closes the cylinder arbor.

  h. The latch slides longitudinally on the left side of the side plate, and is connected to the latch pin by the latch pin stud, causing it to follow the movement of the latch. The latch pin slides in a hole in the frame, and when the cylinder is swung into the frame, the latch pin, under action of the latch spring, is forced into a recess in the ejector and locks the cylinder in position for firing. The latch spring is contained in a hole in the side plate in the rear of the latch slot. The recoil plate is driven into its recess in the frame and secured therein by slightly upsetting the rim.

■ 17. DETAILED FUNCTIONING OF SMITH AND WESSON REVOLVER, M1917 (fig. 5).—*a*. The lock mechanism is contained in the frame and consists of the hammer, with its stirrup, stirrup pin, strut, strut pin, and strut spring; the trigger, with its pin; the trigger lever, with its pin; the rebound slide, with its pin and spring; the hand, with its pin; the hand lever, with its pin and hand lever spring; the cylinder bolt, with its pin, cylinder bolt plunger, cylinder bolt plunger spring, and cylinder bolt plunger screw; the latch, latch plunger, and latch plunger spring.

  b. The hammer and trigger are pivoted on their respective pins. These pins are screwed in place in the left side of the frame and then upset, and are supported on the right side by holes drilled in the side plate to receive them. The rebound slide is held in position by its pin and spring and the rear end of the trigger lever. The lower end of the mainspring fits into a slot in the frame while its upper end engages the hammer stirrup. The mainspring is stressed by screwing up the strain screw, which bears against the mainspring.

  c. The rebound slide houses the spring and slides on the pin against which the spring presses. When the trigger is released, after firing a shot, the rebound slide spring pressing against the rebound slide pin and against the end of the recess in the rebound slide forces the rebound slide forward. The forward end of the rebound slide, pressing against the trigger lever, forces the trigger lever forward and returns the trigger to its original position. The hand being pivoted to the trigger by its pin is thus brought back to its lowest position. After firing, when the hammer is in its extreme forward position, the lowest projection on the hammer lies in the notch on the front end of the rebound slide. As the rebound slide moves forward, the hammer projection is forced out of the notch and on to the flat surface of the slide in the rear of the notch, thus moving the hammer back to its safety position.

d. The revolver may be used either single action or double action. In firing double action, pressure on the trigger causes its upper edge to engage the hammer strut and raise the hammer until the trigger nose itself actually comes into contact with the hammer. After this, the trigger continues to raise the hammer until the hammer is nearly in its full-cock position, when the hammer will escape from the trigger nose and under action of the mainspring will fall, causing the firing pin to strike the cartridge. In firing single action, the hammer is first pulled back with the thumb until the upper edge of the trigger engages in the full-cock notch in the front end of the lower part of the hammer. Pressure on the trigger will then release the hammer, which, under action of the mainspring, will fall and cause the firing pin to strike the cartridge.

e. The cylinder bolt is pivoted on its pin. This is screwed in the left side of the frame and upset, and on the right side is supported by a hole drilled in the side plate. The cylinder bolt plunger spring pressing upward against the cylinder bolt plunger forces it against the bottom of the cylinder bolt and causes the bolt to project through a slot in the frame ready to enter one of the rectangular cuts in the surface of the cylinder. During the first part of the movement of the trigger in cocking the revolver, the lug on the upper front of the trigger engages in the slot in the cylinder bolt, depresses the bolt, and withdraws the nose of the bolt from the cylinder to permit the rotation thereof. The object of the cylinder bolt is to hold the firing chamber in line with the barrel, and also to prevent the cylinder making more than one-sixth of a revolution at the time of cocking. Accordingly the lug on the trigger releases the cylinder bolt before the hammer falls, thus allowing the bolt to arrest the cylinder.

f. The hand is attached by its pivot to the trigger. As the trigger swings on its pin when the hammer is being cocked, the hand is raised, enters a notch in the ratchet on the ejector, and revolves the cylinder, which is simultaneously freed by the cylinder bolt as above described. It serves with the cyl-

inder bolt, after the bolt is released by the trigger, to maintain
the cylinder in proper position at the time of firing, that is,
when the axis of the chamber containing the cartridge to be
fired coincides with the axis of the bore of the barrel. The
hand lever actuated by the hand lever spring, both of which
are housed within the trigger, presses against the lug on the
hand and insures the engagement of the hand with the
ratchet.

g. The cylinder has six chambers. It rotates upon and is
supported by the central arbor of the crane. The crane fits
into a recess in the frame below the barrel and turns on its
pivot arm, which rotates in a hole in that part of the frame
below the opening for the cylinder. The ejector plunger enters the arbor of the crane and is held in place by the thread
on its rear end which engages the corresponding thread in
the front end of the ejector. The shoulder on the ejector
plunger bears against the ejector collar so that when the
ejector plunger is screwed into its housing in the ejector the
ejector spring is compressed. The ejector plunger terminates
at its front end in the ejector plunger head. The ejector, of
which the ratchet forms a part, consists of a rod and the star-
shaped ejector head which engages the clip to cause ejection of
the shells. It is forged in one piece. The ejector spring sur-
rounds the ejector within the arbor of the crane, the front end
bearing on the ejector collar and the rear end on the shoulder
in the rear end of the cylinder.

h. The latch slides longitudinally in its groove within the
left side of the frame. It is connected to the thumb piece
by the thumb piece nut and is actuated by the latch plunger
and latch plunger spring, each housed within the body of the
latch. The center rod passes through the ejector plunger
and ejector projecting at the rear end of the cylinder to
lock the cylinder in position. The shoulder of the center
rod at its rear end bears against the ejector while the ejector
plunger bears against the center rod spring, keeping the
center rod in place. The front end of the center rod does
not come quite flush with the head of the ejector plunger,
thus allowing a recess in the head of the ejector plunger into
which the locking bolt enters. The latch and the hammer
interengage to form an interlock, which prevents cocking of
the hammer when the cylinder is unlatched, and prevents
unlatching of the cylinder while the hammer is cocked.

Since the latch is forced forward by its spring when the cylinder is swung out of the frame, the revolver cannot be cocked unless the cylinder is in position in the frame and latched.

i. The locking bolt is retained in position in its housing by the locking bolt pin and is actuated by the locking bolt spring. The cylinder when closed is retained in its position in the frame and held securely in proper alinement by the center rod which enters its recess in the frame and by the locking bolt, which enters the recess in the head of the ejector plunger. When opening the cylinder, the thumb piece is pushed forward, thus forcing the latch forward. The nose of the latch pushes the center rod out of its recess. The forward movement of the center rod forces the locking bolt forward, thus releasing the cylinder. When the cylinder is swung to the left, out of the frame, it is maintained in its extreme outward position by the crane stop pin which is forced by the crane stop spring into a small depression drilled in the frame.

j. The frame lug is driven into its recess in the frame and riveted. It serves as a stop to the cylinder when ejection of cartridge cases takes place.

Section V

SPARE PARTS AND ACCESSORIES

■ 18. Spare Parts.—In time, certain parts of the revolver become unserviceable through breakage or wear resulting from continuous usage. For this reason spare parts are provided for replacement purposes. They should be kept clean and lightly oiled to prevent rust. They are divided into two groups: organization spare parts and maintenance spare parts.

a. Organization spare parts.—These are extra parts provided with the revolver for replacement of the parts most likely to fail, for use in making minor repairs, and in general care of the revolver. Sets of spare parts should be kept complete at all times. Whenever a spare part is taken to replace a defective part in the revolver, the defective part should be repaired or a new one substituted in the spare parts set as soon as possible. The allowance of these spare parts is prescribed in SNL B–7.

b. Maintenance spare parts.—These are sets of parts provided for the use of ordnance maintenance companies and include all parts necessary to repair the revolver. The allowance of maintenance spare parts is prescribed in the addendum to SNL B–7.

■ 19. Accessories.—The names or general characteristics of many of the accessories required with the revolver indicate their use and application. They consist of the holster, lanyard, cartridge clips, and pistol cleaning kit; and for post, camp, or station issue, arm lockers and arm racks. The pistol cleaning kit contains cleaning brushes and rods, pistol screw drivers, an oiler, and a small brass can in which the set of spare parts is carried.

Section VI

AMMUNITION

■ 20. General.—The information in this section pertaining to the ammunition authorized for use in the revolver, caliber .45, M1917, includes a description of the cartridges, means of identification, care, use, and ballistic data.

■ 21. Classification.—The types of ammunition provided for this revolver are—
   a. Ball, for use against personnel and light matériel targets.
   b. Dummy, for training (cartridges are inert).
   c. Blank, for training.

■ 22. Lot Number.—When ammunition is manufactured, an ammunition lot number which becomes an essential part of the marking is assigned in accordance with pertinent specifications. This lot number is marked on all packing containers and on the identification card inclosed in each packing box. It is required for all purposes of record, including grading and use, reports on condition, functioning, and accidents in which the ammunition might be involved. Only those lots of grades appropriate for the weapon will be fired. Since it is impractical to mark the ammunition lot number on each individual cartridge, every effort will be made to maintain the ammunition lot number with the cartridges once they are removed from their original packing. Cartridges which have been removed from the original packing and for which the ammunition lot number has been lost are placed in grade 3.

It is therefore obvious that when cartridges are removed from their original packings they should be so marked that the ammunition lot number is preserved.

■ 23. GRADE.—AR 775–10 provides for the order in which lots and grades of ammunition are to be used. Ordnance Field Service Bulletin No. 3–5 lists numerically every lot of small-arms ammunition with its correct grade as established by the office of the Chief of Ordnance. Only lots of proper grade will be fired. Grade 3 indicates unserviceable ammunition which will not be fired.

■ 24. IDENTIFICATION.—*a. Markings.*—The contents of original boxes are readily identified by the markings on the box. Similar markings on the carton label identify the contents of each carton.

b. Color bands.—Color bands painted on the sides and ends of the packing boxes further identify the various types of ammunition. The following color bands are used:

```
Cartridge, ball _____ Red
Cartridge, dummy _____ Green
Cartridge, blank_____ Blue
```

c. Types and models.—One model of caliber .45 ball cartridge, one model of caliber .45 blank cartridge, and one model of caliber .45 dummy cartridge are authorized for use in the caliber .45 revolver. These cartridges are designated—
 (1) Cartridge, ball, caliber .45, M1911.
 (2) Cartridge, dummy, caliber .45, M1921.
 (3) Cartridge, blank, caliber .45, M1.
The dummy cartridge is distinguished by its cartridge case which is tinned and has a $\frac{1}{8}$-inch hole in the body. The blank cartridge is characterized by the absence of a bullet. It is provided with a rim which permits it to be fired in and extracted from the revolver without clips.

■ 25. CARE, HANDLING, AND PRESERVATION.—*a.* Small-arms ammunition as compared with other types is not dangerous to handle. However, care must be observed to keep the boxes from becoming broken or damaged. All broken boxes must be immediately repaired and careful attention given so that all markings are transferred to the new parts of the box. The metal liner should be air-tested and sealed if equipment for this work is available.

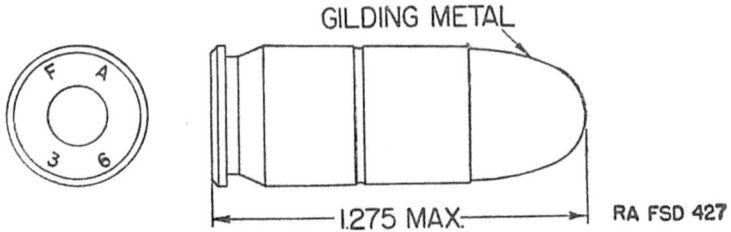

FIGURE 7.—Cartridge, ball, caliber .45, M1911.

b. Ammunition boxes should not be opened until the ammunition is required for use. Ammunition removed from the airtight container, particularly in damp climates, is apt to corrode, thereby causing the ammunition to become unserviceable.

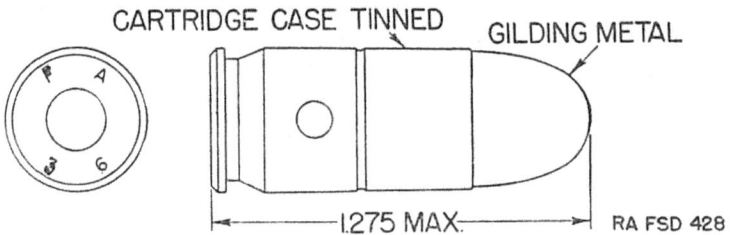

FIGURE 8.—Cartridge, dummy, caliber .45, M1921.

c. Carefully protect the ammunition from mud, sand, dirt, and water. If it gets wet or dirty wipe it off at once. If verdigris or light corrosion forms on cartridges it should be wiped off. However, cartridges should not be polished to make them look better or brighter.

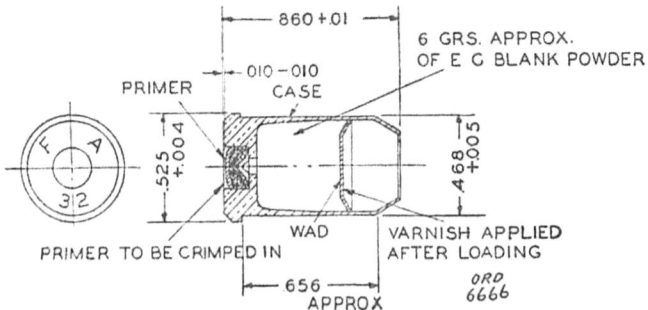

FIGURE 9.—Cartridge, blank, caliber .45, M1.

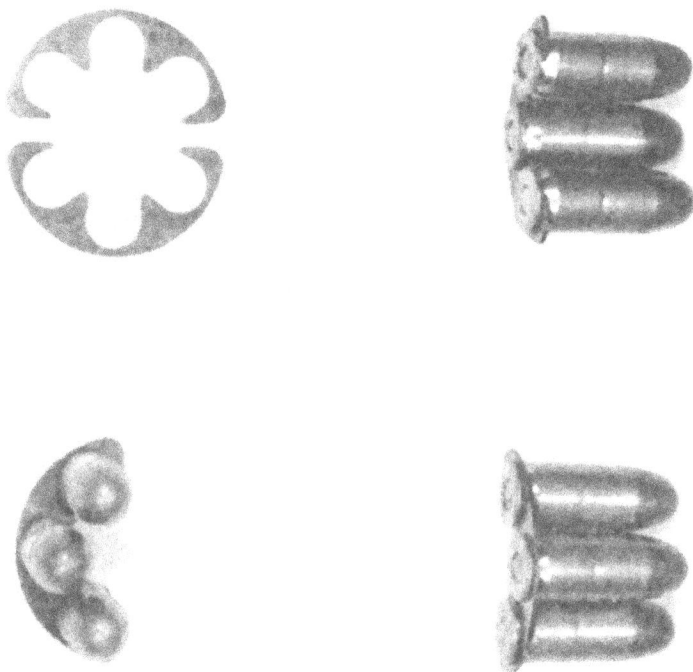

FIGURE 10.—Revolver clip.

d. The use of oil or grease on cartridges is dangerous and is prohibited.

e. Do not fire dented cartridges, cartridges with loose bullets, or otherwise defective rounds.

f. Do not allow the ammunition to be exposed to the direct rays of the sun for any length of time. This is likely to affect seriously its firing qualities.

g. No caliber .45 ammunition will be fired until it has been positively identified by ammunition lot number and grade as published in the latest revision or change to Ordnance Field Service Bulletin No. 3–5.

■ 26. STORAGE.—*a.* Whenever practicable, small-arms ammunition should be stored under cover. Should it become necessary to leave small-arms ammunition in the open it should be raised on dunnage at least 6 inches from the ground and the pile covered with a double thickness of paulin. Suitable trenches should be dug to prevent water flowing under the pile.

b. Fire hazard.—If fired into or placed in a fire, small-arms ammunition does not explode violently. There are small individual explosions of each cartridge, the case flying in one direction and the bullet in another. In case of fire it is advisable to keep those not engaged in fighting the fire at least 200 yards from the fire and have them lie on the ground. It is unlikely that the bullets and cases will fly over 200 yards.

■ 27. BALLISTIC DATA.—*a.* Average velocity of ball ammunition at 25 feet from muzzle, 800 feet per second.

b. Approximate maximum range, 1,600 yards.

SECTION VII

INDIVIDUAL SAFETY PRECAUTIONS

■ 28. RULES FOR SAFETY.—Before ball ammunition is issued, the soldier must know the essential rules for safety with the revolver. The following rules are taught as soon as the recruit is sufficiently familiar with the revolver to understand them. They should be enforced by constant repetition and coaching until their observance becomes the soldier's fixed habit when handling the revolver. When units carrying the revolver are first formed, the officer or noncommissioned officer in charge causes the men to execute INSPECTION PISTOL.

a. Execute UNLOAD every time the revolver is picked up for any purpose. Never trust your memory. Consider every revolver as loaded until you have proved it otherwise.

b. Always unload the revolver if it is to be left where someone else may handle it.

c. Always point the revolver up when snapping it after examination. Keep the hammer fully down when the revolver is not loaded.

d. Never place the finger within the trigger guard until you intend to fire or to snap for practice.

e. Never point the revolver at anyone you do not intend to shoot, nor in a direction where an accidental discharge may do harm. On the range, do not snap for practice while standing back of the firing line.

f. Before loading, open the cylinder and look through the bore to see that it is free from obstruction.

g. On the range, do not load the revolver until the time for firing.

h. Never turn around at the firing point while you hold a loaded revolver in your hand, because by so doing you may point it at the man firing alongside of you.

i. On the range, do not cock the revolver until immediate use is anticipated. If there is any delay, lower the hammer and recock it only when ready to fire.

j. If the revolver fails to fire, open the cylinder and unload if the hammer is down. If the hammer is cocked or partly cocked, a breakage has occurred. In this case hold the revolver at RAISE PISTOL and announce the fact to the officer in charge.

k. To remove a cartridge not fired, open the cylinder and eject, first lowering the hammer if cocked.

l. In campaign, the revolver is carried in the holster fully loaded with the hammer down. The cocked revolver should never be put in the holster whether or not it is loaded.

m. The safety device should be tested frequently.

■ 29. TEST OF SAFETY DEVICES.—*a. Safety.*—With the revolver unloaded and cylinder closed, cock the hammer. Holding the hammer back with the thumb, press the trigger and let the hammer move forward about $\frac{1}{4}$ inch, still holding with the thumb. Release the trigger. Then release the hammer and let it fly forward. If the firing pin projects through the hole in the frame, the safety is faulty.

b. Cylinder bolt.—With the hammer down attempt to rotate the cylinder. If more than about $\frac{1}{64}$ inch in rotation is possible, the cylinder bolt is faulty. Repeat this test with the hammer fully cocked.

NOTE.—With the hammer about one-fourth cocked the cylinder rotates freely.

CHAPTER 2

MANUAL OF THE PISTOL, LOADING AND FIRING, DISMOUNTED

■ 30. GENERAL.—*a.* The movements herein described differ in purpose from the manual of arms for the rifle in that they are not designed to be executed in exact unison. Furthermore, with only a few exceptions, there is no real necessity for their simultaneous execution. They are not, therefore, planned as a disciplinary drill to be executed in cadence with snap and prescision, but merely as simple, quick, and safe methods of handling the revolver. Commands are prescribed only for such movements as may be occasionally executed simultaneously by the squad or larger unit.

b. The revolver is used as a substitute for the automatic pistol. For this reason and in the interests of simplicity the term *pistol* is used in all commands.

c. In general, movements begin and end at the position of RAISE PISTOL.

d. Commands for firing, when required, are limited to COMMENCE FIRING and CEASE FIRING.

e. Officers and enlisted men armed with the revolver remain at the position of attention during the manual of arms. except when their units are presented to their commanders or are presented during ceremonies, at retreat, and at guard mounting. In such cases they execute the hand salute at the command of execution ARMS of 1. PRESENT, 2. ARMS, and resume the position of attention at the command of execution of the next command.

f. When the lanyard is used, it should be of such length that the arm may be fully extended without constraint.

■ 31. To RAISE PISTOL (fig. 11).—The commands are: 1. RAISE, 2. PISTOL. At the command PISTOL unbutton the flap of the holster with the right hand then turn the back of the hand inward and grasp the stock. Draw the revolver from the holster; reverse it, muzzle up, the thumb and last three fingers holding the stock, the forefinger extended outside the trigger guard, the barrel of the revolver to the rear and inclined to the front at an angle of 30°, the hand as high as, and 6 inches in front of, the point of the right shoulder. This is the position of RAISE PISTOL.

■ 32. To LOAD.—Being at RAISE PISTOL the command is: LOAD.
At this command raise the left hand to the front until the forearm is horizontal, palm up.   Place the revolver at the cylinder
in the left hand, latch up, barrel inclined to the left front and
downward at an angle of about 30°.   Press the latch with the
right thumb, push the cylinder out with the second finger of
the left hand, and, if necessary, eject the empty shells by pressing the ejector rod head with the left thumb, right hand
steadying the revolver at the stock.   Take cartridges either
singly or in clips from the belt with the right hand and insert
one in each chamber to be loaded.   Close the cylinder with the
left thumb and resume the position of RAISE PISTOL.

NOTE.—If cartridge clips are not used, empty shells must be
removed from the chambers with the fingernails.

FIGURE 11.—Position of RAISE PISTOL.

33

FIGURE 11.—Position of RAISE PISTOL—Continued.

■ 33. To UNLOAD.—Being at RAISE PISTOL the command is: UNLOAD. Lower the pistol to the left hand and proceed as in paragraph 32, returning unfired cartridges to the belt.

■ 34. To INSPECT REVOLVER (fig. 12).—The commands are: 1. INSPECTION, 2. PISTOL. At the command PISTOL, assume the position of RAISE PISTOL if not already in that position. Open the cylinder by operating the latch with the right thumb and pushing the cylinder to the left with the right forefinger. After the revolver has been inspected, or at the command: 1. RETURN, 2. PISTOL, close the cylinder with the tip of the right thumb.

■ 35. To RETURN PISTOL.—The commands are: 1. RETURN, 2. PISTOL. At the command PISTOL, lower the revolver to the

FIGURE 12.—Position of INSPECTION PISTOL.

holster, reversing it, muzzle down, back of the hand to the body; raise the flap of the holster with the right thumb; insert the pistol in the holster and thrust it home; button the flap of the holster with the right hand.

■ 36. To FIRE—FULLY LOADED WITH BALL AMMUNITION.—*a. Single action.*—Cock the revolver with the right thumb and squeeze the trigger for each shot.

b. Double action.—Executed as in *a* above except that the revolver is cocked by pressing steadily on the trigger.

c. Partially loaded cylinder.—(1) If one or more of the chambers are empty, the cylinder should be rotated so that a loaded chamber will be moved into line with the barrel when the revolver is cocked. With the hammer of the Colt revolver down, the first loaded chamber should be next on the left of the chamber alined with the barrel, since the cylinder rotates clockwise. With the hammer of the Smith and Wesson revolver down, the first loaded chamber should be next on the right of the chamber alined with the barrel, since the cylinder of the Smith and Wesson revolver rotates counterclockwise.

(2) The closed cylinder may be rotated to its proper position by holding the hammer back at about one-fourth full cock.

CHAPTER 3

KNOWN-DISTANCE TARGETS, DISMOUNTED

SECTION I

PREPARATORY TRAINING

■ 37. INSTRUCTION AND PRACTICE.—*a. Relative value.*—(1) Revolver firing is a purely mechanical operation that any man who is physically and mentally fit to be a soldier can learn to do well if properly instructed. The methods of instruction must be the same as are used in teaching any mechanical operation. The soldier must be taught the various steps in their proper order and must be carefully watched and corrected whenever he makes a mistake.

(2) Good shooting is more the result of careful instruction than of mere practice. Unless properly instructed, men instinctively do the wrong thing in firing the revolver. They instinctively jerk the trigger which is the cause of flinching. Hence, mere practice fixes the instinctive bad habits.

(3) If, however, a man has been first thoroughly instructed in the mechanism of correct shooting and is then carefully and properly coached when he begins firing, correct shooting habits rapidly become fixed.

(4) The ultimate object of the training is to develop the ability to fire one or more accurate shots quickly, but training must begin with carefully coached slow fire to attain accuracy and be followed by practice that will gradually shorten the time without sacrificing the accuracy.

b. Methods of instruction.—(1) Revolver instruction is divided into two phases, preparatory instruction and range firing. In the preparatory instruction the soldier learns practically all the principles of good shooting. In range firing he cultivates the will power to apply these principles when using ball ammunition until proper, fixed habits have been acquired.

(2) The principles of good shooting are simple and easy to learn except the trigger squeeze, which is difficult to apply to a loaded revolver. To this important item most of the instructor's time is devoted during the period of range practice.

(3) The six distinct steps in the preparatory instruction are—

(a) Aiming exercises.

(b) Position exercises.

(c) Trigger-squeeze exercises.

(d) Rapid-fire exercises.

(e) Quick-fire exercises.

(f) Examination on preparatory work.

(4) The steps are progressive and must always be taught in proper sequence.

(5) Each of the first five steps begins with a talk by the instructor and a demonstration by a squad which the instructor puts through the exercises that are to constitute the day's work. He shows how the corporal organizes the work in the squad so that no man is idle and how the members of each pair coach one another when they are not under instruction by an officer or a noncommissioned officer. He shows exactly how to execute each of the exercises about to be taken up and explains its purpose and application in revolver shooting.

(6) The instructor who gives these very essential talks and demonstrations may be the organization commander, or he may be a specially qualified officer who has been detailed as instructor. But the actual application of the demonstrated exercises to the men of the command must be by the officers and noncommissioned officers of the organization undergoing instruction.

(7) Instruction must be thorough and must be individual. General instruction of groups of men is not enough. The instructors must see that each man understands each and every point and can apply it.

(8) In peacetime training and in war, when time is available for a complete course of instruction and practice, the form shown in paragraph 38c (which should be explained in the first talk) must be kept by each squad leader and by each platoon leader independently. This form shows at a glance just how much each man knows about each feature of train-

ing and permits concentration of instruction where most needed.

(9) Interest and enthusiasm must be sustained and everything possible should be done to stimulate them. If the exercises are carried out in a manner approximately correct and as a routine piece of work, results will be very disappointing.

(10) It is of utmost importance that the trigger squeeze be explained in such a manner as to give the soldier a clear understanding of how it should be executed.

(11) All authorities on shooting agree that the trigger must be squeezed with a steady increase of pressure. If a man knows when his revolver will go off it is because he suddenly gives the trigger all the pressure necessary. Conversely, if the increase of pressure is steady the man cannot know when the revolver will be discharged. Hence, he is instructed to *squeeze the trigger in such a way as not to know just when the hammer will fall.* This does not mean that the process is necessarily a slow one and that it will take a comparatively long time to fire a shot. Through training, a man can reduce the time used in pressing the trigger to as brief a period as 1 second and still press it in such a manner that he does not know just what part of the second the discharge will take place. When the soldier has acquired the ability to squeeze the trigger properly, even though it be very slowly, he soon learns to shorten the time without changing the process.

(12) Whenever a man is in a firing position, whether it be a preparatory instruction or during practice firing, he must have a coach beside him to watch him and point out his errors.

(13) None of the preparatory exercises are executed by command or in unison by a group of men. Instruction is individual at all times. The men are placed in pairs and alternate in coaching each other. This method gives each man the necessary physical rest without halting the progress of his instruction. He is learning while watching another man and attempting to correct his mistakes.

(14) A great deal of preparatory practice is necessary in order to strengthen the muscles of the hand and arm and to fix the habit of correct trigger squeeze. The periods of exercise should not ordinarily be of long duration. Three or four 10-minute periods per day for a month will produce good results on the range. These periods of instruction can

often be held during waits when troops are on maneuvers or field exercises. Some kind of a mark can always be found that will serve as an aiming point.

(15) It is a good plan to have full-sized pistol targets placed in the vicinity of the barracks to encourage the men to spend part of their time in preparatory practice.

(16) The preparatory exercises should be held out of doors with full-sized pistol targets, but during inclement weather they can be held indoors, using miniature targets, with good results.

c. *Scope of preparatory instruction.*—(1) Each man's revolver is closely examined for defects before the beginning of the preparatory instruction.

(2) Every man who is to fire on the range should be put through the preparatory course. Part of the preparatory instruction may have escaped the men the previous year and part of it has certainly been forgotten; in any case it is beneficial to go over it anew and refresh the mind on the subject.

(3) In peace, noncommissioned officers should be put through a rigid test before the period of preparatory instruction for the organization begins. This is also desirable in war when time is available.

■ 38. First Step: Aiming.—*a. Apparatus required.*—The apparatus required for a set of equipment is listed below. When an entire squad is engaged in this work there should be two sets of this equipment in order that a number of the men do not remain idle. The work of the squad can then be carried on as in rifle marksmanship.

> One sighting bar (fig. 13).
> One revolver rest (fig. 14).
> Two small aiming disks.
> One 5-inch aiming disk.
> Two small boxes, with paper tacked on one side.
> One piece of paper at least 2 feet square and tacked on a wall or frame.

Note.—Men who have once been instructed in the aiming exercises, either in preparation for rifle, pistol, or for revolver firing, will require very little instruction in aiming during subsequent seasons. They will, however, go through the aiming exercises at least once to verify their knowledge of this subject and to assign them a mark in the proper column on the form shown in c below.

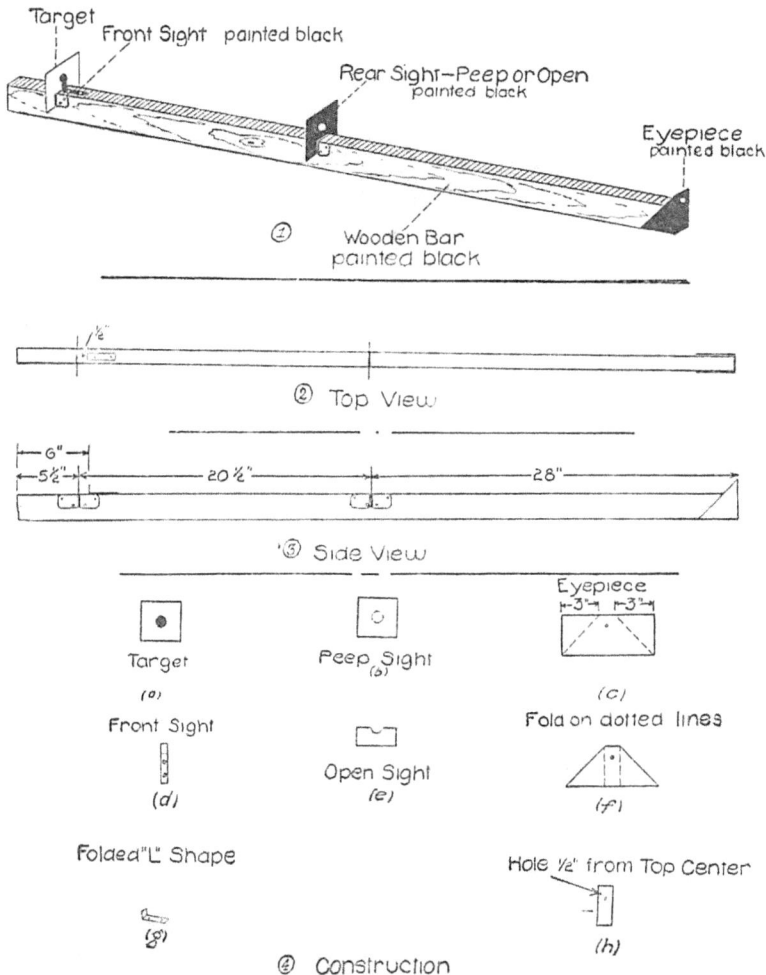

FIGURE 13.—Construction of sighting bar.

Wooden bar—1 by 2 inches by 4 feet 6 inches (approximate).
Eyepiece—Thin metal, 3 by 7 inches; hole, 0.03-inch diameter.
Rear sight—Thin metal or cardboard, 1½ by 3 inches; semicircular
  notch in upper edge, ¾-inch diameter.
Front sight—Thin metal, ½ by 3 inches, bent L shape.
Target—Thin metal or cardboard, 3 by 3 inches, painted white—
  black bull's-eye, ¾-inch diameter in center.
Slits—1 inch deep, may be lined with thin metal strips.

  (1) *Sighting bar.*—(a) The sighting bar is illustrated in
figure 13.

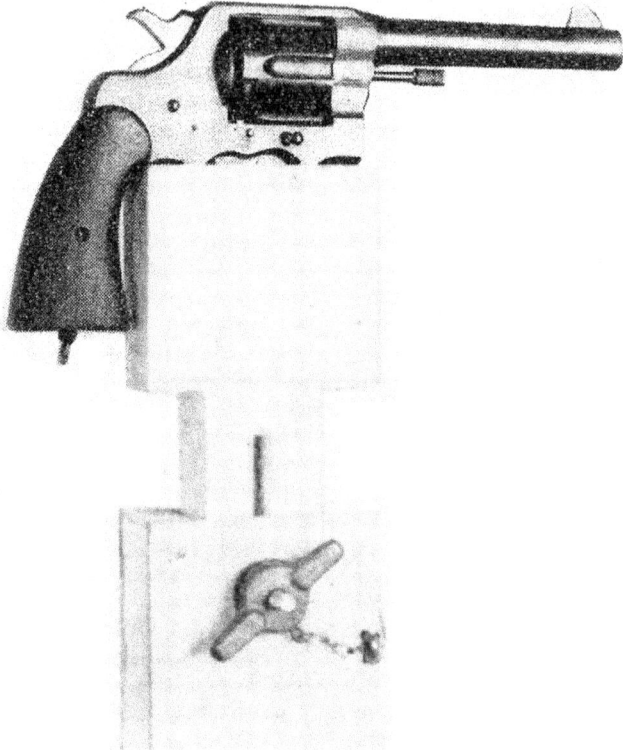

FIGURE 14.—Revolver rest.

(b) Carefully blacken all pieces of tin or cardboard and the top of the bar. Nail the bar to a box about 1 foot high and place the box on the ground, table, or other suitable place.

(c) The sighting bar is used in instruction for two reasons: the sights are larger than on the revolver and errors in aiming can be seen more easily and pointed out to the beginner; the eyepiece of the sighting bar forces the man under instruction to place his eye so that he sees the sights in proper alinement and thus he learns how to aline properly the sights of the revolver. Without an eyepiece the instructor cannot know whether or not the recruit has his eye in proper position.

(2) *Revolver rest.*—(a) To construct a sighting rest for the revolver (fig. 14), build a frame of wood so that the revolver is supported therein. Screw or nail this frame to the top of a post or other object so that the revolver when in position is approximately horizontal. A suitable sighting rest for the revolver may be easily improvised by cutting an additional notch to hold the revolver in one end of the box used as a rifle rest.

(b) Having first learned the principles of aiming by means of the sighting bar, the soldier is taught to apply them to the revolver on its rest.

(3) *Aiming disks.*—(a) For each sighting bar and each revolver rest, a small disk (3 inches in diameter) is made of white cardboard or of tin with white paper pasted on it and with a small bull's-eye in the center. In the exact center of the bull's-eye is a small hole just large enough to admit the point of a pencil. For indoor or close-range work the bull's-eye should not be larger than a 50-cent piece.

(b) There should be one 5-inch aiming disk for each squad for shot-group exercise at 25 yards. The large disk should be of tin, painted black, with a handle 4 or 5 feet long and of the same color as the paper on which the shot groups are to be made.

b. *Sighting exercises.*—(1) *First exercise.*—(a) The squad leader or instructor shows a sighting bar to his squad or group and points out the front and rear sights, the eyepiece, and the removable target. He explains the use of the sighting bar as follows:

1. The front and rear sights on the sighting bar represent enlarged revolver sights.
2. The sighting bar is used in the first sighting exercise because with it small errors can be seen easily and explained to the pupil.
3. The eyepiece requires the pupil to place his eye in such position that he sees the sights in exactly the same alinement as seen by the coach.
4. There is no eyepiece on the revolver, but the pupil learns by use of the sighting bar how to aline the sights properly when using the revolver.
5. The removable target attached to the end of the sighting bar is a simple method of readily alining the sights on a bull's-eye.

(b) The instructor, using the open sight, adjusts the sights of the sighting bar with target removed to illustrate a correct alinement of the sights. He has each man of the assembled group look through the eyepiece at each of the sight adjustments.

(c) The instructor adjusts the sights of the sighting bar with various small errors in sight alinement and has each man of the assembled group endeavor to detect the errors.

(d) The instructor describes a correct aim, again showing the illustration to each man (fig. 15). He explains that the top of the front sight is seen through the middle of the open sight and is raised to a height so that its top is level with the outside edges of the open sight and just touches the bottom of the bull's-eye so that all of the bull's-eye can be clearly seen.

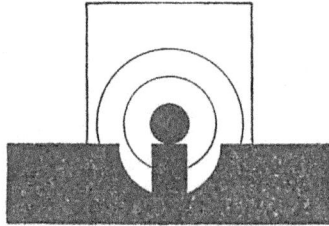

FIGURE 15.—Normal open sight.

(e) The instructor explains that the eye should be focused on the bull's-eye in aiming, and he assures himself by questioning the pupils that each man understands what is meant by focusing the eye on the bull's-eye.

(f) The instructor adjusts the sights of the sighting bar and the removable target so as to illustrate a correct aim and has each man of the group look through the eyepiece to observe this correct aim.

(g) The instructor adjusts the sights and the removable target of the sighting bar so as to illustrate various small errors and has each man in the group attempt to detect the error.

(h) The exercise described above having been completed by the squad leader or other instructor, the men are placed in pairs and the exercise is repeated by the coach-and-pupil method.

(2) Second exercise.—(a) With the revolver on the revolver rest and the sights pointing at a blank sheet of paper

on a board or on the wall, stand with the head in the same relative position as in firing the revolver and look through the sights (fig. 16). Then by signal or by word have the disk moved until the bottom edge of the bull's-eye is in exact alinement with the sights. Then command HOLD and move away from the revolver and let the man undergoing instruction look through the sights to see the proper aim.

FIGURE 16.—Sighting exercise.

(b) Have the man under instruction look through the sights while he directs the disk to be moved until the sights are alined on the bottom of the bull's-eye. The instructor then looks through the sights to see if any error has been made.

(c) Have the sights adjusted on the bull's-eye with various very slight errors and see if the man under instruction can detect them readily.

(3) *Third exercise.*—Using the sighting rest for the revolver, require the man under instruction to direct the marker to move the disk until the sights are aimed at the bottom edge of the bull's-eye and to command HOLD. The instructor then looks at the aim, and after noticing whether

45

the aim is right or wrong commands: MARK. The marker, without moving the disk, makes a pencil mark on the paper through the hole in the center of the bull's-eye. Repeat the operation until three marks have been made. The instructor looks at the aim each time, but he says nothing to the man until all three marks have been made and joined together so as to make a shot group. The faults, if any, are then pointed out. The size and shape of the shot group are discussed and the exercise is repeated several times. At 30 feet, using the small bull's-eye, the shot group should be small enough to be covered by a dime.

c. *Progress.*—This form is used during the period of preparatory instruction. Its object is to show at all times the state of instruction of each man and to insure his thorough instruction in all necessary points before range practice begins.

Names	Recruit instruction				Marksmanship—dismounted										Remarks
	Functioning and operation	Safety precautions	Test of safety devices	Care and cleaning	Sighting bar	Exercises with the revolver rest	Holding the breath	Position of the hand	Position exercise	Trigger squeeze exercise	Instruction in calling the shot	Rapid-fire exercise	Quick-fire exercise	Ability as a coach	Final examination	

Method of marking:

```
  X
          Fair.
```
```
  X
    X         Good.
```
```
  X
    X
      X       Very good.
```
```
  X
    X  X
  X    X      Excellent.
```
```
  X  X
    X
  X  X         Excellent. Has in-
               structional ability.
```

■ 39. SECOND STEP: POSITION.—*a. Essentials of proper position.*—To assume the proper position for firing it is necessary to know how to aim, how to grasp the revolver, how to hold the breath properly, and the correct position of the body with relation to the target.

(1) *How to grasp the revolver.*—(*a*) To take the grip, hold the pistol in the left hand and force the rear surface of the grip back into the crotch formed between the thumb and forefinger of the right hand so that the hammer just clears the web between the thumb and forefinger. The thumb is carried parallel with or slightly higher than the forefinger; it should never be lower. Close the three lower fingers on the stock firmly but not with a tense grip (fig. 17). The little finger may be placed under the butt.

FIGURE 17.—How to grasp revolver.

(*b*) The thumb and forefinger squeeze the frame of the revolver, but the ball of the thumb does not always touch the revolver, depending on the conformation of the man's hand. By this pressure, movement to the right or left is controlled, and the trigger squeeze can be better applied and coordinated.

(*c*) The right arm is extended; the muscles of the arm are firm without being rigid. There should be no bending of the arm at the elbow when the revolver is fired. On the other hand, the arm should not be locked at the elbow. When the firer is shooting properly, after recoil the revolver arm should automatically carry the revolver back to the position shown in figure 18.

(2) *How to hold the breath.*—(a) The proper method of holding the breath is important because without instruction many men hold the breath in the wrong way or do not hold it at all.

(b) To hold the breath, draw into the lungs a little more air than an ordinary breath, let a little of the air out and stop the rest by closing the throat. Do not hold the breath with the throat open or by the muscular effort of the diaphragm.

(3) *Position of body* (fig. 18).—(a) The position of the body is a little more than half faced to the left, the feet 12 to 18 inches apart, depending on the man, the head erect,

FIGURE 18.—Position of body.

and the body perfectly balanced when the revolver is held in the shooting position.

(b) The whole position should be natural and comfortable. Upon assuming the position there is some point to which the revolver points naturally and without effort. If this point is not the center of the target, the whole body must be shifted so as to bring the target into proper alinement. Otherwise the firer will be firing under a strain because he will be pulling the revolver on the target by muscular effort for each shot. Any unnecessary tensing of any of the muscles of the hand, arm, or body will cause tremors and should therefore be avoided.

b. *Position exercise.*—(1) Required for this exercise: A line of L targets with firing points at 15 and 25 yards, or a line of small aiming bull's-eyes placed at the height of the shoulder.

(2) The men, armed with the revolver, are placed in one line at 1-pace intervals. Give the command: 1. INSPECTION. 2. PISTOL, and verify the fact that all revolvers are unloaded.

(3) Demonstrate the position of the hand in gripping the stock and describe the grip in detail.

(4) Require each man to grip the stock of his revolver in the prescribed manner, using the free hand to grasp the barrel and set the stock well back in the revolver hand between the thumb and the first finger.

(5) Describe the correct method of holding the breath while aiming and require each man to practice it a few times.

(6) Demonstrate the correct position of the whole body when firing, explaining in detail the position of the feet, legs, body muscles, arms, and head.

(7) Require each man to assume the correct firing position. The officers and noncommissioned officers of the organization correct individuals who are at fault.

(8) The above exercises having been completed, instruction becomes individual under a coach. The men are placed in pairs opposite L targets or opposite small aiming bull's-eyes and take turns coaching each other.

(9) The details of the position exercises are—

(a) Grasp the stock with the correct grip.

(b) Face target, then face half left.

(c) Separate the feet 12 to 18 inches.

(*d*) Aline the sights on the bottom edge of the bull's-eye, arm extended.

(*e*) Hold the breath.

(*f*) As soon as the arm becomes tired or the aim becomes unsteady, assume the position of RAISE PISTOL.

(*g*) The revolver should be removed from the right hand and the muscles of the hand, arm, and shoulder relaxed and exercised before resuming the grip. This should also be done between shots in slow fire.

(10) After the firer has completed the position exercise he may repeat it with a weight, such as a pair of field glasses in a case suspended from the right arm. The weight is suspended first between shoulder and elbow, then from the forearm, then from the wrist, and finally from the barrel of the revolver, interspersed with short rests. The value of this exercise lies in developing the muscles of the shoulder and arm.

(11) (*a*) The hammer is not raised during the position exercises and the trigger is grasped very lightly with the finger.

(*b*) After a short rest repeat the exercise.

(*c*) The man acting as coach watches carefully and corrects all errors.

(*d*) The man under instruction and the coach change places as the officer in charge of the instruction desires. This should be every 3 to 5 minutes.

(*e*) Only a few hours in all should be devoted to the position exercises, as all of its details are included in the trigger-squeeze exercise.

■ 40. THIRD STEP: TRIGGER SQUEEZE.—*a. Importance of correct trigger squeeze.*—(1) The recruit can readily learn to aim and hold the aim either on the bull's-eye or very close to it for at least 10 seconds. When he has learned to press the trigger in such a manner as not to spoil his hold he becomes a good shot. All men flinch in firing the revolver if they know the exact instant at which the discharge is to take place. This is an involuntary action which cannot be controlled. A sudden pressure of the trigger may derange the aim slightly, but the extreme inaccuracy of a shot fired in this way is due mainly to the flinch, that is, the thrusting forward of the hand to meet the shock of recoil. Any man who holds the sights of the revolver as nearly on the bull's-eye as possible and continues to press on the trigger with a uniformly increasing pressure until the revolver goes off is a *good shot*. Any man who has learned to increase the pressure on the trigger only when the sights are in alinement with the bull's-eye, who holds the pressure when the muzzle swerves, and who continues with the pressure when the sights are again in line with the bull's-eye is an *excellent shot*. Any man who tries to "catch his sights" as they touch the bull's-eye and to set the revolver off at that instant is a *very bad shot*.

(2) The apparent unsteadiness of the revolver while being held on the bull's-eye does not cause much variation in the striking place of the bullet due to the fact that the movement is of the whole extended arm and revolver. But the flinch which always accompanies the sudden pressure of the trigger deflects the muzzle of the revolver and causes the bullet to strike far from the mark. In squeezing the trigger the pressure must be *straight* to the rear. There is a tendency on the part of some men to press the trigger also to the left.

b. Calling the shot.—To call the shot is to state where the sights were pointed at the instant the hammer fell; thus. high, a little low, to the left, slightly to the right, bull's-eye, etc. If the soldier cannot call his shot correctly in range practice he did not press the trigger properly and consequently did not know where the sights were pointed when the hammer fell.

c. Trigger-squeeze exercises.—(1) *First exercise.*—(a) Required for this exercise: A line of L targets with a firing point at 25 yards.

(b) Give the command: 1. INSPECTION, 2. PISTOL. and verify the fact that all revolvers are unloaded.

(c) The squad leader explains to his squad the details of this exercise which are—

1. Cock the piece.
2. Take the correct grip.
3. Take the correct position.
4. Aline the sights on the target and start the squeeze, gradually increasing the pressure on the trigger until the hammer falls.
5. Call the shot.

51

6. Rest the hand.

7. Repeat the above operations, firing double action.

NOTE.—In squeezing the trigger in firing double action, the firer should squeeze rapidly until the hammer is almost back to the full cock position. He then continues the squeeze as explained above. Repeated practice accustoms the firer to the feel of the trigger in this method of firing.

(*d*) The squad leader assures himself that all the men understand the details of this exercise. The work is then carried on by pairs working together, coach and pupil. Members of the squad should change over frequently to avoid tiring the muscles of the arm. Extended trigger-squeeze exercise is necessary and the periods should be short but frequent.

(*e*) The duties of the coach are to—

1. See that the firer takes the correct grip.
2. See that the firer takes a correct position.
3. Watch the hand of the firer to see that he is gradually increasing the pressure on the trigger.
4. See that the firer rests his shooting hand after the hammer falls.
5. See that the firer calls the shot each time.

(2) *Second exercise.*—(*a*) Give the command: 1. INSPECTION, 2. PISTOL, and verify the fact that all revolvers are unloaded.

(*b*) The squad leader explains to his squad the details of this exercise, which are—

1. Cock the piece.
2. Take the correct grip.
3. Take the correct position.
4. Aline the sights on the target and start the squeeze. Close the eyes and continue to squeeze until the hammer falls.
5. When the hammer falls, open the eyes and check the aim to see if it has been deranged.
6. Repeat the above exercise, firing double action.

NOTE.—The firer should be able to keep on the target. If he is off persistently, he should check on his grip and position to see that they are correct.

(*c*) The duties of the coach in this exercise are the same as in the first trigger-squeeze exercise.

■ 41. FOURTH STEP: RAPID FIRE.—*a. When taken up.*—Training for rapid fire should be taken up after the principles of slow fire, particularly the trigger squeeze, are thoroughly understood, and some facility in applying them has been gained.

b. General principles.—(1) Rapid fire is the same as slow fire except that the piece remains pointed at the target for five consecutive shots, and there is no pause or delay between the discharge of one shot and the application of the operations to fire the next shot.

(2) To attain the degree of accuracy required for proficiency in rapid fire, the revolver must be cocked by use of the thumb and not by using double action.

(3) Basically, the grip on the stock (except for a slightly tighter feel and the fact that the finger tips now contact the stock), the position of the arm, and the position of the body are the same as for slow fire.

(4) It is important that a uniform grip be maintained throughout the firing of the score. Any shift in the position of the revolver in the hand or variation in the pressure of the hand and fingers during the firing of the score results in inaccuracy. A great deal of practice "dry shooting" is necessary to develop skill.

(5) To fire the first shot, the revolver is brought from the position of RAISE PISTOL by the shortest route to the firing position with the sights properly alined on the aiming point. This is done by a smooth but deliberate and rapid extension of the right arm straight from the shoulder, cocking the piece, inserting the forefinger in the trigger guard during the movement, and holding the breath.

(6) In rapid fire, time is gained by—

(a) Taking position accurately.

(b) Applying a heavy initial pressure on the trigger as soon as the sights are alined and then maintaining a continuously increasing pressure until the shot is fired.

(c) Rapid and smooth cocking of the piece.

(d) Keeping the focus of the eye on the bull's-eye during the firing of the entire string.

(e) Absorbing the shock of recoil at the shoulder, not at the wrist or elbow, thereby reducing the movement of the gun to a minimum.

Caution: Every effort should be made to overcome the disastrous tendency to save time by pulling the trigger quickly when the aim is perfect.

c. Cocking revolver.—(1) *General.*—There are two methods of cocking the revolver for the succeeding shots, the side method and the straight-back method. Both methods are good. Each requires a great deal of practice before sufficient skill is acquired to cock the hammer without shifting the position of the stock in the hand. The method to be used by the individual depends upon the size, shape, and muscular development of the hand. For that reason both methods should be taught and practiced until it is determined which one proves the more satisfactory. Thereafter only that method should be practiced.

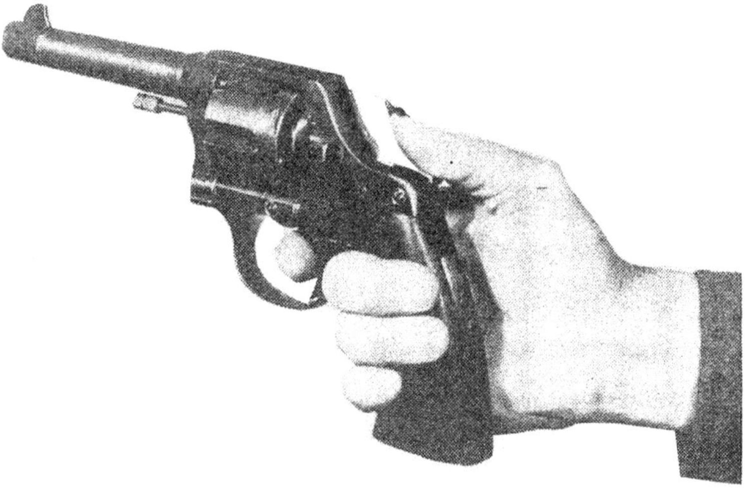

FIGURE 19.—Side method of cocking revolver.

(2) *Side method* (fig. 19).—(a) The recoil of the revolver causes it to rise about 4 to 6 inches above the point of aim. As it reaches the top of the upward movement, relax the grip slightly and place the ball of the thumb on the spur of the hammer. Exert a downward pressure with the thumb and at the same time move the muzzle with a wrist motion to the right about 4 inches. This combined action of the thumb and movement of the muzzle causes the hammer to snap to full cock.

(b) During this movement it is important that the fingers on the left side of the stock be kept in place to assist in controlling the revolver.

54

(c) Immediately after the hammer has snapped to the full cock position, the revolver is moved back into aiming position and the thumb replaced along the side of the frame.

(d) During the operation of cocking and bringing the revolver back to the aiming position, the muzzle must be kept elevated so that the front sight is visible and can be readily alined in the rear sight notch. This obviates the possibility of losing time in hunting for the front sight and in alining it in the rear sight notch which often occurs when the muzzle is allowed to sag.

(e) Some individuals may experience difficulty in keeping the grip in the same position on the revolver during the firing of a string. In most cases the hand tends to work higher on the stock thus restricting the action of the thumb in cocking, and making it necessary to regrasp the revolver in the middle of the string. This difficulty may be overcome by altering the grip slightly so that the little finger is placed under the bottom of the butt of the stock (fig. 17).

(3) *Straight-back method* (fig. 20).—(a) With this method of cocking the revolver, the grip is not loosened nor is the revolver shifted from its line of recoil. As soon as the shot is fired and while the gun is in recoil, the thumb is placed on the hammer spur and the hammer drawn straight back to the full cock position by the action of the thumb only. During the time the hammer is being drawn back, the revolver is lowered from its uppermost recoil position to the aiming position. As soon as the hammer is cocked, the thumb is replaced alongside the frame.

(b) This method is simpler and has the following advantages over the side method:

1. Permits the grip to be more uniformly maintained throughout the firing of the rapid-fire string.
2. Since there is no side movement of the revolver during the process of cocking, the sights can be more readily realined and brought back on the point of aim.

(c) The straight back method has the disadvantage, however, that many men are unable to flex their thumbs sufficiently to draw the hammer all the way back. This causes the thumb to become cramped when the hammer is about two-thirds of the way to full cock, and necessitates regrip-

①

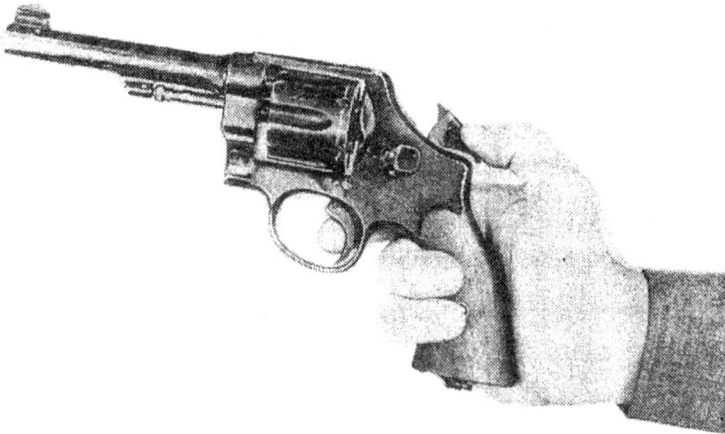

②

FIGURE 20.—Straight-back method of cocking revolver.

ping the piece to complete the cocking of the hammer. This is a very bad feature as it results in a loss of time and cadence in firing the string. This difficulty may, in some cases, be obviated by slightly lowering the grip. However, care should be exercised that the grip is not too low, as this results in the stock sliding further into the hand with each shot fired.

Men who experience this difficulty should be required to use the side method of cocking the hammer.

d. Cocking exercise.—(1) *General.*—This exercise is held for the purpose of acquiring a smooth and rapid cocking operation.

(2) Practice in cocking the revolver should be conducted employing both of the above-described methods in order to enable the soldier to determine which method is best adapted to the conformation and flexibility of his hand. Thereafter, only the method selected is practiced. Before being considered proficient, the soldier must be able to cock the revolver at least 15 times in 10 seconds. The first hour of rapid-fire training should be devoted to cocking exercises. Thereafter, each pupil should be given additional practice from time to time until he is considered proficient.

(3) *Procedure.*—The exercise is conducted by the coach-and-pupil method. The instructor explains and demonstrates both methods. Emphasis is placed on the following points: The importance of maintaining a uniform grip; the necessity of keeping the muzzle of the revolver high; and the advantages of keeping the eye focused on the front sight. This exercise should not be continued longer than about 10 seconds at a time. Frequent changes of coach and pupil are necessary to prevent undue tiring of the muscles of the arm and hand. After requiring the pupil and his coach to take position on the line, the instructor commands: 1. COCKING EXERCISE—READY, 2. EXERCISE, 3. CEASE FIRING, 4. REST. At the first command the pupil grasps the grip as described in paragraph 39 and extends the revolver to the firing position. At the second command the eye is focused on the front sight, without attempting to aim, and the trigger is squeezed causing the hammer to fall. The operations of cocking the hammer by one of the prescribed methods are then continued until CEASE FIRING is given, when the revolver is brought to the position of RAISE PISTOL. At the command REST, the revolver is returned to RAISE PISTOL while the hand and arm are rested.

(4) *Duties of the coach.*—In the cocking exercise the coach insures that—

(*a*) The grip is properly taken.

(b) The muzzle is kept elevated during the cocking of the hammer.

(c) The eye is kept focused on the front sight.

(d) The trigger is allowed to move to its forward position immediately after the hammer falls.

(e) In the side method, the same grip is taken after each cocking operation; the fingers are kept in place on the stock; and the muzzle is moved to the right by the action of the wrist and not by moving the arm.

(f) In the straight-back method, the grip, except for the action of the thumb, is not shifted or loosened; the hammer is drawn back by the action of the thumb only; and the revolver remains in the same plane during the cocking operation.

e. *Rapid-fire exercise.*—(1) Required for this exercise: A row of L targets or a row of aiming bull's-eyes.

(2) Give the command: 1. INSPECTION, 2. PISTOL, and verify the fact that all revolvers are unloaded.

(3) Explain to the assembled command that the trigger squeeze is the same in rapid fire as in slow fire.

(4) Demonstrate the correct method of bringing the revolver by the shortest route to the aiming position. Show how this is done from RAISE PISTOL and in drawing the revolver from the holster in an emergency.

(5) Show how the revolver is held to facilitate rapid cocking with the right thumb without disarranging the hold.

(6) Demonstrate the method of cocking the revolver with the right thumb, without bending the elbow.

(7) Show how the revolver is kept as nearly on the mark as possible during the whole score. Caution the men to avoid unnecessary flourishes or movements between shots.

(8) Demonstrate—

(a) The action of the revolver in recoil when a shot is fired.

(b) How the arm should not be permitted to bend at the elbow.

(c) How the revolver should move upward through a small arc and be deflected from the original point of aim only a short distance.

(d) How the forefinger should move forward after the explosion only far enough to allow the trigger to become reengaged with the hammer and immediately after cocking start pressing the trigger for the next shot.

(e) How the eye should not be allowed to close when the explosion occurs.

(f) How the breath should be held for each shot.

(9) (a) The above demonstrations having been completed, the men are placed in front of the line of targets in pairs, one to practice and one to coach. The exercise is then carried on exactly the same as rapid fire in range practice. If a line of disappearing targets has been arranged for this exercise, the targets appear, remain in sight the allotted time, and then disappear. While the targets are in sight, each man undergoing instruction attempts to fire five shots (simulated fire), cocking the piece for each shot except the first with the thumb.

(b) If the targets are stationary the exercise begins with the command: 1. COMMENCE. 2. FIRING, and ends with the command: 1. CEASE, 2. FIRING.

(c) After each three or four scores of simulated fire the men of each pair are directed to change places, the firer becoming the coach and the coach becoming the firer.

(10) (a) In this exercise the coach carefully watches the man and corrects all errors in grip, position, trigger squeeze, cocking and manipulation of the piece, paying particular attention to the trigger squeeze.

(b) Rapid-fire exercises should be frequent but not of long duration.

(c) It is advisable to extend the time limits several seconds when rapid-fire exercise is first taken up. The time limit is then gradually reduced until it corresponds to the time prescribed for range firing, record practice.

■ 42. FIFTH STEP: QUICK FIRE.—a. Training for quick fire.— (1) The training for quick fire is taken up after the rapid-fire exercise has been practiced sufficiently to be understood thoroughly. Thereafter, exercises in slow fire, rapid fire, and quick fire should all be continued until the end of the period of preparatory training.

(2) For each shot the revolver is brought from RAISE PISTOL to the aiming position by the shortest route after the target appears.

(3) The revolver is cocked with the thumb after each shot in this exercise before the position of RAISE PISTOL is resumed.

b. Quick-fire exercise.—(1) Required for this exercise: A line of E targets that can be operated as bobbing targets from a pit or screen, or a line of E targets so arranged on pivots that the edge can be turned toward the firer when the target is not exposed.

(2) Give the command: 1. INSPECTION, 2. PISTOL, and verify the fact that all revolvers are unloaded.

(3) Explain to the assembled command that the trigger squeeze is the same in quick fire as in slow fire.

(4) Demonstrate the correct method of bringing the piece from RAISE PISTOL to the aiming position.

(5) Show how the revolver is cocked between shots.

(6) The above demonstrations having been completed, the men are placed in pairs in front of the line of bobbing targets, one man of each pair to act as coach for the other man. The exercise is then carried on exactly the same as quick fire in range practice. The targets appear, remain in sight the allotted time, and then disappear. After the targets appear each man undergoing instruction brings his revolver from RAISE PISTOL to the aiming position, aims, fires one shot (simulated fire), recocks his piece, and returns to the position of RAISE PISTOL. After three or four scores of simulated fire the men of each pair are directed to change places.

(7) The coach watches carefully the man going through the exercises and corrects all errors in the grip, position, holding the breath, trigger squeeze, cocking and the manipulation of the piece, paying particular attention to the trigger squeeze. It is advisable to extend the time limit about 2 seconds for each shot when quick-firing exercise is first taken up. The time is then gradually reduced until it corresponds to the time prescribed for range firing, record practice.

(8) When disappearing targets cannot be provided for this exercise it may be held with stationary E targets. The command UP is given to signify that the targets are in sight, and the command DOWN to signify that they have been withdrawn.

(9) Practice in quick fire should be held frequently, but the periods of practice should not be of long duration.

(10) If the range is some distance from the area designated for preparatory exercises, or it is impracticable to arrange for a line of bobbing targets, L targets may be substituted for the bobbing targets.

■ 43. SIXTH STEP: EXAMINATION.—At the completion of the preparatory instruction, the instructor should assure himself by an examination that every man understands thoroughly and can explain every phase of the preparatory training. The questions and answers given below are merely examples. Each man should be required to explain each item in his own words.

Instructor: Examine your revolver to see that it is unloaded.

Q. What are the safety devices on the revolver? A. The safety and the cylinder bolt.

Q. Show me how you test the safety. A. I cock the revolver, hold the hammer back with the thumb while pressing the trigger to disengage it from the hammer, let the hammer down a little way, release the trigger, then release the hammer. I see if the hammer falls all the way forward.

Q. Show me how you test the cylinder bolt. A. With the hammer down I attempt to rotate the cylinder. If it moves more than about $\frac{1}{64}$ inch in either direction the revolver is faulty.

Q. If the tests of the safety devices fail at any time, what should you do? A. I should report the matter at once to my platoon or company commander.

Q. Show me how to load the Colt revolver with three (five) loose rounds so that it will fire the first time it is cocked and each time thereafter. A. I open the cylinder and insert three (five) cartridges in consecutive chambers. I then close the cylinder, pull back the hammer to about one-fourth full cock position and rotate the cylinder until the first loaded chamber is next on the *right* of the empty chamber alined with the barrel.

Q. Show me how you let the hammer down on a loaded revolver without firing. A. I pull the hammer a little to the rear of full cock with the thumb and holding it back I press the trigger, let the hammer forward about ¼ inch with the thumb, release the trigger, and then lower the hammer all the way with the thumb.

Q. What is this (indicating a sighting bar)? A. A sighting bar.

Q. What is it used for? A. To teach men how to aim.

Q. Why is it better than a revolver for this purpose? A. Because the sights are much larger and slight errors can be seen more easily and pointed out.

Q. What does this represent? A. The front sight.

61

Q. What does this represent? *A.* The rear sight.

Q. What is this? *A.* The eyepiece.

Q. What is it for? *A.* To make the man hold his head in the right place so that he will see the sights properly alined.

Q. Is there an eyepiece on the revolver? *A.* No. A man learns by the sighting bar how the sights look when properly alined, and he must hold the revolver while aiming so as to see the sights in the same way.

Q. Adjust the sights of this sighting bar so that they are in proper alinement with each other. (Verified by instructor.)

Q. Now that the sights are properly adjusted, have the small bull's-eye moved until the sights are aimed at it properly. (Verified by instructor.)

Q. Tell me what is wrong with this aim. (The instructor now adjusts the sights of the sighting bar on the bull's-eye with various very slight errors, requiring the man to point out the error.)

Q. Show me how you grip the stock of the revolver.

Q. Show me the position you take when you are going to shoot.

Q. How do you squeeze the trigger? *A.* I squeeze it with such a steady increase of pressure as not to know exactly when the hammer will fall.

Q. If the sights get slightly out of alinement while you are squeezing the trigger, what do you do? *A.* I hold the pressure I have on the trigger and go on with the increase of pressure only when the sights become alined again.

Q. If you do this can your shot be a bad one? *A.* No.

Q. Why? *A.* Because I cannot flinch, for I do not know when to flinch, and the sights will always be lined up with the bull's-eye when the shot is fired because I never increase the pressure on the trigger except when the sights are properly alined.

Q. When you are practicing in slow fire and your arm becomes unsteady and your aim uncertain, what should you do? *A.* I should come back to RAISE PISTOL without firing the shot and then try again after a short rest.

Q. If it is impossible for you to hold the revolver steady, can you still do good shooting? *A.* Yes; if I press the trigger properly.

Q. Tell me why that is. *A.* Because the natural unsteadiness of the arm moves the whole revolver and the barrel remains nearly parallel to the line of sight. But if I give the trigger a sudden pressure the front end of the barrel will be thrown out of line with the target, and the bullets will strike far out from the mark.

Q. What causes this deflection of one end of the revolver when the trigger is given a sudden pressure? A. The sudden pressure itself causes some of it, but most of it is caused by the flinch that always accompanies this kind of a trigger pressure.

Q. What does a man do when he flinches in shooting a revolver? A. He usually thrusts his hand forward as if trying to meet the shock by suddenly stiffening all his muscles.

Q. Must the trigger always be squeezed slowly in order to do it correctly? A. No. I squeeze it the same way in rapid fire and quick fire. The time is shorter but the process is the same.

Q. What is meant by calling the shot? A. To say where you think the bullet will hit as soon as you shoot and before the shot is marked.

Q. How can you do this? A. By noticing exactly where the sights point at the time the revolver is fired.

Q. If a man cannot call his shot correctly, what does it indicate? A. That he did not squeeze the trigger properly and consequently did not know where the sights were pointed at the instant the discharge took place.

Q. Show me how you hold your breath while aiming.

Q. Take your revolver. Aim at that bull's-eye and squeeze the trigger a few times, calling the shot each time. (The instructor particularly notes the holding of the breath.)

Q. Show me how you come to a position of AIM from RAISE PISTOL.

Q. Show me how you come to the aiming position in drawing the revolver from the holster in an emergency.

Q. Take this revolver and fire a rapid fire score at that target (simulated fire). I will command COMMENCE FIRING to start the score and CEASE FIRING to stop it.

Q. Fire a score (simulated fire) at that quick-fire target. I will give the command UP when it is supposed to come into sight, and the command DOWN when it is supposed to be withdrawn from view.

Q. What do you do in case a cartridge misses fire in combat? A. I recock the pistol for the next shot.

Q. Are there any points about revolver firing that you do not understand?

NOTE.—In all the demonstrations by the man undergoing examination the instructor carefully notes all points that are covered in the preparatory exercises. Each man is put through a thorough test along the line indicated in these questions and answers before he is allowed to fire.

SECTION II

COURSES TO BE FIRED

■ 44. GENERAL.—AR 775–10 prescribes details as to who will fire and ammunition allowances.

■ 45. INSTRUCTION PRACTICE.—The following tables prescribe the firing in instruction practice in the order followed by the individual soldier. Target L is used in much of the practice as the bull's-eye makes competition keener and shows up errors as no other target can.

a. Slow fire.

TABLE I.—*Slow fire—target L*

Range	Time	Scores (5 shots each), minimum
15 yards	No time limit	1
25 yards	do	1

Unlimited time is permitted for slow fire in order to permit proper explanation of the causes of errors and indication of corresponding remedies. It is intended to be the elementary phase of instruction in the proper manipulation of the weapon and for determining and correcting the personal errors of the firer.

b. Rapid fire.

TABLE II.—*Rapid fire—target L*

Range	Time	Scores (5 shots each), minimum
15 yards	15 seconds per score	1
25 yards	20 seconds per score	1

If pits are used, time is taken at the pits as in rapid-fire rifle practice. If pits are not used, time is taken at the firing point. The target being up, the soldier stands with the loaded weapon at RAISE PISTOL. The command READY is given and the soldier cocks his revolver and resumes the position of RAISE PISTOL. The command: 1. COMMENCE, 2. FIRING, is given and the soldier must fire one score within the prescribed limit of time, at the end of which the command: 1. CEASE, 2. FIRING, is given. Intervals of time are measured from the last words of the commands.

c. *Quick fire.*

TABLE III.—*Quick fire—target E—bobbing*

Range	Time	Scores (5 shots each), minimum
15 yards	3 seconds per shot	2
25 yards	3 seconds per shot	2

■ 46. RECORD PRACTICE.—The following tables prescribe the firing in record practice in the order followed by the individual soldier. The procedure is as in instruction practice.

a. *Slow fire.*

TABLE IV.—*Slow fire—target L*

Range	Time	Scores (5 shots each)
25 yards	No time limit	2

b. *Rapid fire.*

TABLE V.—*Rapid fire—target L*

Range	Time	Scores (5 shots each)
15 yards	15 seconds per score	2
25 yards	20 seconds per score	2

c. Quick fire.

TABLE VI.—*Quick fire—target E—bobbing*

Range	Time	Scores (5 shots each)
25 yards	3 seconds per shot	3

SECTION III

CONDUCT OF RANGE PRACTICE

■ 47. COACHING METHODS.—*a. Range practice.*—(1) The object of range practice is to teach the men to apply with a loaded revolver the principles of good shooting that they have learned during the preparatory exercises.

(2) Each man while firing must have a coach to correct him whenever he violates any of these principles.

(3) Slow-fire practice should be carried on until the man under instruction thoroughly understands the principles of good shooting.

(4) When rapid fire and quick fire are first taken up, the time limit should be extended a few seconds. The time should then be gradually reduced until the scores are being fired in the time prescribed for record practice.

b. Dummy cartridges.—(1) Dummy cartridges are of great value in teaching both slow and rapid fire.

(2) *Dummy cartridges must not be used except on the firing line of the pistol range.* The same precautions are observed as in using service ammunition.

c. Slow fire.—(1) The coach stands on the left side of the firer in such a position as to be able to observe the latter's trigger finger, his grip, his eye, and his position. It is the duty of the coach to correct all errors. The coach fills the clips for the firer and hands them to him. At the beginning of range practice the clips should be filled partly with service ammunition and partly with dummy cartridges, in which case the coach loads the revolver. The firer must not know how many dummy cartridges are in the clip or the order in which they are loaded.

(2) The object of using dummy cartridges is to show the coach whether or not the man under instruction is squeezing

the trigger correctly, and in case of an improper trigger squeeze to bring the fact forcibly to the attention of the firer himself. When a loaded cartridge is fired, the flinch is often masked by the recoil of the revolver and the firer is not conscious of having flinched. When the hammer falls on a dummy cartridge which the firer thinks is loaded, the sudden stiffening of the muscles and the thrusting forward of the hand to meet the shock that does not come are apparent to everybody in the vicinity, including the firer himself. The mixing of dummy cartridges with service ammunition causes the man to make a determined effort to press the trigger properly for all shots.

(3) The firing of scores with dummy cartridges and service ammunition should not be confined to the early stages of training. It is advisable to have some practice of this kind each day during the entire period of instruction practice.

(4) The following items of instruction are given to a pupil on beginning range practice, even though that person has already done a great deal of shooting. Once a person has been put through this instruction it is usually not necessary to repeat it during subsequent periods of range practice.

(a) Explain the method of grasping the piece.

(b) Show the amount of force used in gripping the stock by grasping the pupil's hand, saying: "This is too tight a grip" (gripping his hand very tightly). "This is too loose a grip" (gripping his hand loosely). "This is the right amount of force to use in gripping the stock" (gripping his hand with the firm but comfortable grip that should be used in shooting).

(c) Explain and demonstrate the position of the body, the feet, and the arm (par. 39 and fig. 18), and have the pupil assume this position.

(d) Explain the proper method of aiming.

(e) Explain that any man can aim and hold well enough for a good score. Have the pupil assume the proper position and aim at the target with an empty revolver without attempting to press the trigger to see how long he can hold the sights on or near the bull's-eye. Explain to him that this aiming at the target with an empty gun demonstrates how near to the center his bullets will strike provided he presses the trigger properly.

(f) Explain the proper method of pressing the trigger.

(g) Have the pupil aim at the target with an empty revolver and then press the trigger for him several times as described in d below (fig. 21), directing the pupil to call the shot each time the hammer falls.

(h) Have the pupil aim at the target with a loaded revolver and then press the trigger for him as described in d below, directing him to call the shot each time the piece is fired. Fire a score of five shots in this way.

(i) Have the pupil fire a score of five shots, pressing the trigger himself to see if he can press the trigger properly and make as good a score as the one made when the coach pressed the trigger.

FIGURE 21.—Coach pressing trigger.

d. *Squeezing trigger.*—One method of showing the men under instruction how to squeeze the trigger properly is to have him hold and aim the revolver while the coach presses the trigger. This is done in the following manner:

(1) The coach demonstrates the value of correct trigger press to the student by placing his hands in the position shown in figure 21 and pressing on the end of the pupil's trigger finger with his left thumb. The coach cautions the pupil neither to assist nor resist the pressure which is put on the end of his trigger finger but to devote his whole attention to his aim and hold.

(2) The coach must be careful to apply a slow, steady pressure to the finger of the pupil and at the same time not interfere with the pupil's aim while applying this pressure. As a rule, the coach should consume from 5 to 10 seconds in putting sufficient pressure on the pupil's finger to fire the revolver.

(3) When pressing the trigger for a pupil as above described, the coach should hold his head well to the rear to keep from having his left ear too near the muzzle of the piece.

(4) If the firer shows a tendency to apply the last part of the squeeze himself by giving the trigger a sudden pressure, he is directed to remove his finger from the trigger guard, and the coach applies the pressure directly to the trigger instead of through the finger of the man under instruction.

e. Calling the shot.—Men should be required to call each shot in slow fire. If a man does not call the shot immediately after firing, the coach directs him to do so.

f. Coaching rapid fire.—(1) The firing of scores with dummy cartridges and service ammunition mixed is a very valuable form of rapid-fire practice. The coach loads the revolver in such a way that the firer cannot know the order in which the cartridges are placed.

(2) The coach must watch the man closely, and each time he is seen to flinch, whether on a loaded or a dummy cartridge, the coach should caution him.

(3) When the hammer falls on a dummy cartridge the firer cocks the revolver and continues the exercise.

g. Coaching quick fire.—(1) The use of dummy cartridges and the coaching methods are the same in quick fire as in rapid fire.

(2) The occasional use of dummy cartridges in both rapid fire and quick fire should be continued throughout the entire period of instruction practice.

■ 48. SAFETY PRECAUTIONS ON THE RANGE.—*a.* Never place a cartridge in the chamber of the revolver until you have taken your place on the firing line and received the command: LOAD.

b. Always open the cylinder and eject all cartridges before you leave the firing point.

c. Always hold the loaded revolver at the position of RAISE PISTOL except while aiming and cocking.

d. When firing ceases temporarily, hold the piece at RAISE PISTOL. Do not assume any position except RAISE PISTOL without first unloading.

e. If one or more cartridges remain unfired at the end of a rapid-fire or quick-fire score, unload immediately.

■ 49. RANGE ORGANIZATION.—*a.* The work on the range should be so organized that no man is idle for any length of time. A good arrangement is four or six orders per target. It should never be necessary to assign more than six orders. If there is not a sufficient number of targets to provide for this, the extra men should remain off the range and be given other instruction.

b. One method is to have a line of pistol targets on a flank of each firing point of the rifle range so arranged that the firing points of the rifle range and of the pistol range are on one line. There should be about 50 yards interval between the rifle range and the pistol range. The targets may be placed on the ground instead of in pits. The bobbing targets are arranged to revolve on their own axis and are operated from behind the firing line by means of cords. When the targets are to be marked the whole line ceases firing, unloads revolvers, and moves up to the targets to record the hits and paste the shot holes. In slow fire the coach can keep the firer informed as to the location of his hits by use of field glasses.

c. When the time is short and range facilities and proper supervision permit, rifle firing and dismounted revolver firing may be carried on at the same time. While one group is firing on the rifle range the other is firing on the pistol range. As the men complete a score with the rifle they move to the pistol range and their places at the rifle firing point are filled by men who have completed a score of revolver firing. As soon as all men present have completed their scores with the rifle, the whole group moves back to the next firing point (moving the pistol targets if necessary) and continue as before with the alternate rifle and revolver firing.

d. The pistol targets may be placed so that the line of fire is at right angles to the line of fire of the rifle range if the terrain permits. When it is not practicable to have revolver firing and rifle firing at the same time, other means should be adopted to keep the men occupied while they are not actually firing or coaching.

■ 50. TARGET DETAILS.—The personnel for the supervision, operation, and scoring of targets during recorded firing consists of officers, noncommissioned officers, and privates as follows:

a. One commissioned officer assigned to each four targets. The officer will take up and sign each duplicate score card as soon as a complete score is recorded.

b. One noncommissioned officer assigned to each target to enter scores on the duplicate score cards and to direct and supervise the detail pasting the target. This noncommissioned officer will be selected, except at a one-company post, from an organization other than the one firing on the target which he supervises. When the post is garrisoned by a single company so that it is impossible to detail noncommissioned officers of other companies to supervise the marking and scoring, those duties are performed by the noncommissioned officers of the firing company.

c. One or two privates to operate, mark, and paste each target.

d. When the targets are not placed in pits, target details may be reduced to one commissioned officer for each four targets, one noncommissioned officer for not to exceed each two targets, and one private to each target.

e. The noncommissioned officer examines the target after each score is fired and enters the score on the score card, initialing same. He directs the private to paste the target after the score is recorded and marked and examines the target to see that no shot holes are left unpasted.

■ 51. REGULATIONS GOVERNING RECORD PRACTICE.—*a. Coaching prohibited.*—Coaching of any nature after the firer takes his place on the firing point is prohibited. No person may render or attempt to render the firer any assistance whatever while he is taking his position or after he has taken his position at the firing point. Each firer must observe the location of his own hits.

b. Shelter for firer.—Sheds or shelter for the firer are not permitted on any range.

c. Cleaning.—Cleaning is permitted only between scores.

d. Gloves.—A glove may be worn on either or both hands.

e. Pieces loaded on command.—Pieces are not loaded except by command or until position for firing has been taken.

f. Shots cutting edge of bull's-eye or line.—Any shot cutting the edge of the figure or bull's-eye is signaled and recorded as a hit in the figure or the bull's-eye. Because the limiting line of each division of the target is the outer edge of the line separating it from the exterior division, a shot touching this line is signaled and recorded as a hit in the higher division.

g. Slow fire score interrupted.—If a slow fire score is interrupted through no fault of the person firing, the unfired shots necessary to complete the score are fired at the first opportunity thereafter.

h. Misses.—In all firing before any miss is recorded the target is carefully examined by an officer.

i. Accidental discharges.—All shots fired by the soldier after he has taken his place on the firing line (and it is his turn to fire, the target being ready) are considered in his score even if his piece was not directed toward the target or is accidentally discharged.

j. Firing on wrong target.—Shots fired upon the wrong target are entered as misses upon the score of the man firing no matter what the value of the hits upon the wrong target may be. In rapid fire the soldier at fault is credited with only such hits as he may have made on his own target.

k. Two shots on same target.—In slow fire, if two shots strike a target at the same time or nearly the same time, and if one of these shots was fired from the firing point assigned to that target, the hit having the highest of the two values is entered on the soldier's score and no record is made of the other hit.

l. Withdrawing target prematurely.—In slow fire, if the target is withdrawn from the firing position just as the shot is fired, the scorer at that firing point at once reports the fact to the officer in charge of the scoring on that target. That officer investigates to see if the case is as represented. Being satisfied that such is the case he directs that the shot be not considered and that the man fire another shot.

m. Misfires.—In case of a misfire in rapid fire or quick fire the soldier ceases firing and takes the position of RAISED PISTOL. In rapid fire the score is repeated. In quick fire the score is continued after the defective cartridge has been replaced.

n. Unused cartridges in rapid fire.—Each unfired cartridge is recorded as a miss.

o. Disabled revolver.—If during the firing of a rapid or quick fire score the revolver becomes disabled through no fault of the firer, the procedure outlined in *m* above is followed.

p. More than five shots in rapid fire.—When a target has more than five hits in rapid fire it is not marked unless all the hits have the same value. It is then marked and the firer given that value for each shot he actually fired, not to exceed five.

q. Cocking revolver for first shot in rapid fire.—The firer may cock the revolver for the first shot in rapid fire and quick fire prior to the appearance of the target or before the command COMMENCE FIRING is given.

r. Score cards and scoring.—(1) Entries on all score cards are made in ink or with indelible pencil. No alteration or correction is made on the card except by the organization commander who initials each alteration or correction made.

(2) The scores at each firing point are kept by a non-commissioned officer of some organization other than that firing on the target to which he is assigned, except in case of a one-company garrison when company officers exercise special care to insure correct scoring. As soon as a score is completed the score card is signed by the scorer, taken up and signed by the officer supervising the scoring, and turned over to the organization commander. Except when required for entering new scores on the range, score cards are retained in the personal possession of the organization commander and not allowed in the hands of an enlisted man from the beginning of record practice until the required reports of range practice have been rendered.

(3) In the pit the officer keeps the scores for the targets to which he is assigned. As soon as a score is completed he signs the score card. He turns these cards over to the organization commander at the end of the day's firing or at such times as requested.

(4) Upon completion of record firing and after the qualification order is issued, the pit score cards of each man are attached to his official score card kept at the firing point. These cards are kept available for inspection among the company records for 1 year and then destroyed.

■ 52. COMPUTING SCORES.—The soldier's individual score is computed on a percentage basis. The soldier's percentage in firing each of the tables listed in paragraph 46 is calculated separately, then the sum of these percentages is divided by three to give the final average percentage.

■ 53. CLASSIFICATION.—The individual classification to be attained and the method of determining qualification are as prescribed in AR 775–10. Firers are classified as pistol experts, pistol sharpshooters, pistol marksmen, and unqualified.

SECTION IV

KNOWN-DISTANCE TARGETS AND RANGES; RANGE PRECAUTIONS

■ 54. TARGETS.—*a. Target E.*—Target E is a drab silhouette about the height of a soldier in a kneeling position made of bookbinder's board or other similar material (fig. 22). Hits are valued at 1. Any shot cutting the edge of the target is a hit.

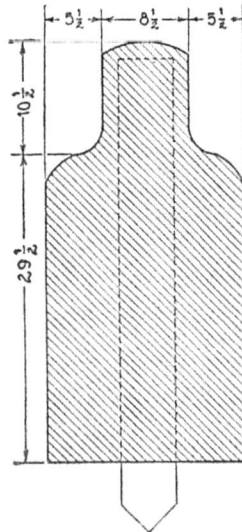

FIGURE 22.—Target E.

b. Target E—bobbing.—Target E—bobbing is so arranged as to be fully exposed to the firer for a limited time, edge of target toward firer when target is not exposed (fig. 23).

c. Target L.—Target L is a rectangle 6 feet high and 4 feet wide, with black circular bull's-eye, 5 inches in diameter and seven outer rings (fig. 24). Value of hits in the bull's-eye, 10. The diameter of each ring and value of hits are as follows:

Diameter	Value of hit
8½ inches	9
12 inches	8
15½ inches	7
19 inches	6
22½ inches	5
26 inches	4
46 inches	3
Outer, remainder of target	2

d. Small-bore targets.—No specific targets are prescribed for small-bore practice with the revolver, and any suitable targets may be used. If the targets specified in paragraph 60 are not available, the following targets issued by the Ordnance Department are suitable and may be used: Rifle—SB–A–2, SB–A–3, and **SB–B–5**.

■ 55. RANGES.—*a. General.*—Class A target ranges for the rifle as described in FM 23–5 and 23–10 may be used for revolver practice, if available. The pistol target L may be placed on the sliding target carriage for both slow and rapid fire. Bobbing targets are not ordinarily placed in the pits of rifle ranges but are set up nearby. If sufficient space is available, ground other than rifle ranges is used for revolver practice.

b. Rules for selection.—As the nature and extent of the ground available for revolver practice and also the general climatic conditions are often widely dissimilar for different military posts, it is impossible to prescribe any particular rules governing the selection of ranges, but only to express certain general conditions to which ranges should be made to conform.

c. Safety necessary.—For posts situated in thickly settled localities where the extent of military reservation is limited, the first condition to be fulfilled is that of safety for those living or working near or passing by the range. This require-

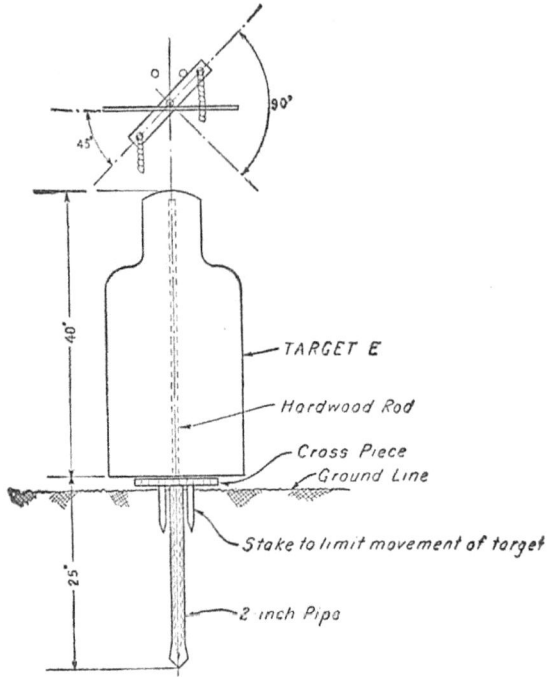

FIGURE 23.—Target E—bobbing.

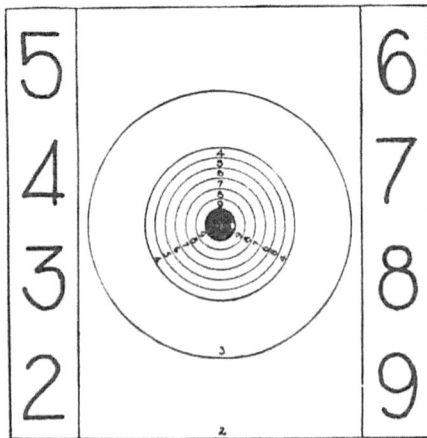

FIGURE 24.—Target L.

ment can be secured by selecting ground where a natural butt is available or by making an artificial butt sufficiently extensive to stop wild shots. For complete safety there should be no road, building, or cultivated ground nearer than 300 yards to either flank of the range nor less than 1,600 yards to its rear.

d. Direction of the range.—If possible a range should be so located that the direction of firing is toward or slightly to the east of north. Such location gives a good light on the face of the targets during the greater part of the day. However, safety and suitable ground are more important than direction.

e. Best ground for range.—Smooth, level ground or ground with only a very moderate slope is best adapted for a range. The target should be on the same level with the firer or only slightly above him. Firing down hill should be avoided.

f. Size of range.—The size of the range is determined by its plan and by the number of troops that will fire on it at a time.

■ 56. PRINCIPLES GOVERNING CONSTRUCTION.—*a. Intervals between targets.*—Intervals between targets are equal to the width of the targets themselves. When the necessity exists for as many targets as possible in a limited space, this interval may be reduced. Bobbing targets should be placed a minimum of 5 yards apart.

b. Protection for markers.—When pits are not used, markers remain in rear of the firing line except during cessation of fire when their duties require them to move to the targets.

c. Artificial butts.—If an artificial butt is constructed as a bullet stop, it should be of earth not less than 30 feet high with a slope of not less than 45°. It should extend about 5 yards beyond the outside targets and should be placed as close behind the targets as possible. The slope should be sodded.

d. Hills as butts.—A natural hill to form an effective butt should have a slope of not less than 45°. If originally more gradual it should be cut into steps, the face of each step having that slope.

e. Number of targets.—Each target should be designated by a number.

f. Measuring the range.—The range should be carefully measured and marked with a stake in front of each target at each firing point. The stakes should be about 12 inches above the ground and painted white. These stakes then designate the firing points for the different targets at the different distances. Particular care should be taken that each stake thus placed is on a line at right angles to the face of its own target.

g. Danger signals.—One or more danger signals are placed near the range to warn passersby when firing is in progress. They should be placed on the roads or on the crest of the hill where they can be plainly seen by those passing.

SECTION V

SMALL-BORE PRACTICE

■ 57. GENERAL.—Where facilities and equipment permit, all soldiers who have satisfactorily passed the examination on preparatory exercises should be advanced to small-bore practice before taking up range practice. The actual firing and observed results stimulate endeavor. Rapid progress, particularly in teaching trigger squeeze, may be made. There is no recoil or loud report to induce nervousness or flinching and the soldier soon learns that he can make good scores if he observes the proper methods and precautions in which he has been instructed. Small-bore practice is not only valuable to the beginner but it affords to the good shot a means of retaining his efficiency throughout the year. Continued practice is essential in maintaining revolver marksmanship.

■ 58. OBJECT.—The object of small-bore practice is to provide a form of marksmanship training with the caliber .22 revolver and ammunition which represents the application of the principles taught in the preparatory exercises. Small-bore practice provides an excellent means of improving the shooting of organizations and sustaining interest in marksmanship throughout the year. Every effort should be made by all organizations to fire the small-bore course prior to the regular marksmanship season. The firing of this course enables the organization commander to visualize the state of training of his command and to concentrate his efforts on the training of those who are most deficient.

■ 59. CONTINUOUS PRACTICE.—Small-bore practice should be carried on throughout the year subject to such limitations as may be imposed by available ammunition and range facilities. All persons who have never been properly instructed in shooting methods prescribed herein should be given a thorough course of preparatory instruction before being permitted to fire on the small-bore range. All small-bore practice is properly organized and supervised in accordance with the methods of instruction as prescribed in this manual.

■ 60. SMALL-BORE PRACTICE COURSE.—Prior to range firing, the minimum number of scores shown in the following small-bore practice table should be fired by each individual required to fire the revolver, dismounted. The procedure of firing is similar to that in range firing. No reports of the results of small-bore practice are required, but the firing record of individuals should be posted in order to stimulate interest and competition among the men of the organizations.

Small-bore practice table

Range	Slow-fire target		Rapid-fire target		Quick-fire target	
	Iron small-bore target or paper target X		Same as slow fire		Target E—bobbing	
	Time	Scores	Time	Scores	Time	Scores
5 yards	No time limit	2	20 seconds per score.	2		
10 yards	do	2	do	2		
15 yards					3 seconds per shot.	2

■ 61. ADDITIONAL PRACTICE.—In addition to the minimum number of scores prescribed in the table above, small-bore practice should be carried on throughout the year, the amount and details of the practice being left to the discretion of the organization commander. Varied targets such as tin cans, bottles, pendulums, and moving targets stimulate interest. Matches between individuals and teams of the same or different units should be promoted.

CHAPTER 4

FIRING AT FIELD TARGETS

■ 62. GENERAL.—*a.* After individual practice is completed, firing at field targets may be commenced by all organizations authorized to fire this practice.

b. This firing is practiced by squads.

■ 63. FIELD FIRING RANGE.—Eight targets E—bobbing are placed approximately in line with 5-yard intervals and with ropes sufficiently long to reach behind the starting line. The starting line is 50 yards from the targets and is marked by a flag at each side of the range. The area should be clear of trees and high brush, level or sloping gently upward toward the targets.

■ 64. FIELD FIRING COURSE.—The following table outlines the course to be fired. The course may be repeated at will as often as is consistent with ammunition allowances.

Formation	Scores	Targets	Range	Pace
Squad of 8 men deployed at 5-yard intervals.	2 scores, each man firing 6 shots in each score.	8 targets E—bobbing.	Between 50 and 15 yards.	Walk.

NOTE.—Hits are valued at 1. Any shot cutting the edge of the target is a hit.

■ 65. METHOD OF FIRING COURSE.—*a.* The squad is formed at the starting line facing the targets with 5-yard intervals. Fully loaded revolvers are in holsters.

b. The officer in charge of firing commands: 1. FORWARD, 2. MARCH. All men move toward the targets at the walk. By signal the officer in charge of firing causes the targets to be exposed for 6 seconds when the line has moved forward about 5 yards.

c. When the targets first appear, the men individually draw and raise the revolver and cock the hammer. They fire as many shots as they desire at each exposure of the target. During the periods when the targets are exposed all men must halt. During periods when targets are not exposed all men move toward the targets at the walk with revolvers at RAISE PISTOL.

d. The officer in charge of firing causes the targets to be exposed 3 seconds after the line of men has advanced each 5 yards, until men reach a line 15 yards from the targets where the targets are exposed for the last time. Thus targets are exposed six times; the first time for 6 seconds and all other times for 3 seconds.

e. Each man must fire only on the target directly in front of him. He may fire by single action or double action.

f. Care must be taken that the line be kept approximately dressed. If a man gets 3 yards in front of or behind the line, he must be cautioned of his error and must correct it immediately.

g. For purposes of safety a noncommissioned officer is stationed in rear of the squad and follows it forward during each run to enforce the provisions of *f* above.

h. Firing ceases on command of the officer in charge of firing. Men stand fast, revolvers are unloaded, and the officer in charge of firing commands: 1. INSPECTION, 2. PISTOL, 3. RETURN, 4. PISTOL.

CHAPTER 5

ADVICE TO INSTRUCTORS

Section I

GENERAL

■ 66. Provisions Not Mandatory.—The information and suggestions contained in this chapter are not mandatory unless so specifically stated. They are furnished as a guide for the personnel responsible for the instruction of troops in the subjects contained herein.

■ 67. Method of Instruction.—The applicatory system of instruction is used for instruction in subjects of the nature found in this manual. This system consists of explanation, demonstration, application (practical work), and examination. (See FM 21–5.)

a. Explanation.—The initial explanation and demonstration of any particular phase of the instruction are presented to the assembled unit by the instructor assisted by essential demonstration personnel. The general purpose of the entire course or period of instruction should be explained first. The various phases or steps of the course should then be presented in a series of explanations and demonstrations.

b. Demonstration.—(1) Demonstrations which are skillfully conceived and executed expedite and simplify instruction as well as stimulate interest. Successful demonstrations are usually short and concise. They leave the student with an exact impression stripped of superfluous details. The demonstrations incident to all subjects should be arranged in progressive sequence, and where practicable should alternate with practical work to permit the student to fix these successive phases of instruction in his mind.

(2) The men who constitute the demonstration unit should be carefully selected for their intelligence, ability, and appearance. They should be thoroughly trained and rehearsed in the duties they are to perform so that the demonstration

will proceed smoothly and illustrate clearly and simply the phase of instruction being presented.

(3) The equipment used for demonstrations should be the best available. A demonstration platform or an area in which the students can be assembled quickly at a position from which they can see and hear every part of the demonstration is essential.

(4) Interest is added and valuable instruction given by repeating demonstrations, including common errors, and requiring the students to detect these errors.

c. *Application (practical work).*—(1) This third step of instruction is of major importance since it gives the student an opportunity to actually accomplish that which has been previously explained and demonstrated.

(2) During the practical work phase of instruction, best results are obtained if the unit is divided into groups. Groups should consist of from four to eight men depending upon the number of men undergoing instruction and the number of assistant instructors available. Each group is provided with a set of equipment and placed under the direct supervision of a trained assistant instructor. The group then executes the prevously demonstrated phase of instruction, individuals rotating within the groups, until all men have mastered the instruction.

(3) The initial allotment of time and equipment should be made carefully. However, the instructor should not hesitate to alter this allotment if the majority of the men fail to master the instruction within the allotted time or are kept at one exercise to the point of boredom. The frequent rotation of duties within each group is preferable to keeping each man in one position for a long time.

d. *Examination.*—An informal oral or practical examination should be conducted upon completion of each phase of instruction. In addition to the required examination before starting range practice, the organization commander should conduct such additional examinations as are necessary to insure that all men have completed the training.

Section II

MECHANICAL TRAINING

■ 68. General.—The entire unit to be instructed is assembled in a suitable area and divided into conveniently sized groups, each under the supervision of an assistant instructor. The instruction is centralized under the supervision of the unit instructor. Explanation and demonstration are concurrent, each assistant instructor demonstrating the elements of the particular phase of instruction as the instructor explains it from the platform. For short periods of practical work the instruction is decentralized under the assistant instructors.

■ 69. DISASSEMBLY AND ASSEMBLY OF REVOLVER.—*a. Equipment required.*—One revolver per man; one pistol cleaning kit per group.

b. Procedure.—(1) Have assistant instructor disassemble and assemble the revolver while the instructor is explaining the procedure.

(2) Assistant instructors explain and demonstrate the procedure, and each student performs each operation in unison with the assistant instructor. When acquainted with the procedure each student disassembles and assembles the revolver without assistance.

(3) Ask questions.

■ 70. CARE AND CLEANING.—*a. Equipment.*—Same as paragraph 69 plus additional for special demonstrations.

b. Procedure.—(1) Explain the need for keeping the revolver clean, lubricated, and in proper condition, comparing it with any other piece of machinery.

(2) Explain the proper method of cleaning, at the same time demonstrating the process.

(3) Instruction in cleaning is carried out throughout the year under supervision of squad and platoon commanders.

(4) Ask questions. Inspect revolvers frequently.

■ 71. FUNCTIONING.—*a. Equipment.*—Same as paragraph 69.

b. Procedure.—(1) Explain the various steps which take place in firing the revolver, both single action and double action.

(2) Ask questions.

■ 72. ACCESSORIES.—*a. Equipment.*—One each of the accessories listed in paragraph 19.

b. Procedure.—(1) Explain and demonstrate the use of each accessory.

(2) Students examine accessories.

(3) Ask questions.

■ 73. INDIVIDUAL SAFETY PRECAUTIONS.—*a.* Equipment.—Same as paragraph 69.

b. Procedure.—(1) Explain and demonstrate the various safety rules listed in paragraph 28. Explain and demonstrate tests for safety devices.

(2) Students practice the application of the various safety rules and make the tests of safety devices. Observation of the rules for safety becomes a habit only after constant practice over a long period. Squad, platoon, and other commanders should be constantly alert to enforce the safety rules at every opportunity.

(3) Ask questions.

SECTION III

MANUAL OF THE PISTOL

■ 74. GENERAL.—*a.* Instruction in the manual of the pistol is carried out concurrently with dismounted drill and with previous instruction in this chapter. Manual of the pistol lends itself well to the applicatory system of instruction.

b. Equipment.—Each man is equipped with a revolver, two cartridge clips, holster, and belt.

c. Procedure.—(1) The instructor explains and demonstrates each movement in the manual of the pistol, employing a trained demonstration unit.

(2) Groups are separated under assistant instructors and are drilled in the various movements.

(3) Each group is tested by the instructor at the end of each period and a critique is conducted.

SECTION IV

MARKSMANSHIP

■ 75. GENERAL.—*a.* Marksmanship is the basic step in training the soldier to employ successfully the revolver in combat. A soldier subconsciously employs in combat the principles he has been taught in marksmanship, hence these principles must be sound.

b. The procedure used in conducting marksmanship instruction is similar to that used in the preceding sections of this chapter except that it is more decentralized. During instruction in preparatory exercises the entire unit is assembled initially under the unit instructor assisted by a trained demonstration unit. Following the initial explanation and demonstration the groups move to their individual sets of equipment and start practical work under the assistant instructors.

c. Firing exercises should be conducted under centralized control.

■ 76. PREPARATORY RANGE TRAINING.—a. *General.*—(1) A thorough course in preparatory range training is essential. During this period the soldier learns all the mechanics of target practice except actual firing. Preparatory training may be done in barracks or other nonfiring areas.

(2) Adequate time should be allowed and thorough supervision provided to insure that each man has thoroughly mastered the instruction before he is permitted to fire.

(3) Each step is taken up in proper order and training in that step completed by each man before the next step is begun. If men fail to progress uniformly, groups should be rearranged so that instruction is not held up by men who are slow to learn.

(4) A careful record of the progress of each man and each group should be kept in order that the instructor will know the progress of instruction and when the men are ready for range practice.

b. *Equipment per group.*—(1) One sighting bar.

(2) One revolver rest.

(3) Two small aiming disks.

(4) One 5-inch aiming disk.

(5) Two small boxes with paper tacked on one side.

(6) One target frame on which is placed a blank sheet of paper at least 2 feet square.

(7) One target L.

(8) One target E—bobbing.

(9) One revolver, cartridge clips, holster, and belt.

(10) Material for blackening sights.

(11) Tissue paper for copying shot groups.

(12) Pencils.

(13) Additional equipment such as blackboard, charts, and drawings as decided by the instructor.

c. *Procedure.*—(1) Each phase of preliminary training is explained and demonstrated when instruction in that phase begins by the unit instructor, employing a trained demonstration group.

(2) Groups are separated and practical work is conducted under the supervision of assistant instructors.

(3) Examination is conducted by asking questions and by observing the results obtained by each man during practical work.

■ 77. FIRING OF COURSE.—*a. General.*—Details of administration and supply are determined to a large extent by the number of men undergoing instruction and the range facilities available. These matters should be anticipated in order that men who are firing will not be distracted. Men who are waiting to fire may be perfecting preliminary instruction or they may derive valuable instruction by watching others fire and by listening to critiques.

b. Equipment.—Instruction in marksmanship is facilitated by having all revolvers in perfect mechanical condition.

■ 78. CONSTRUCTION OF TARGETS AND RANGES.—*a. General.*— For detailed information relative to targets and target accessories see Table of A l l o w a n c e s, Targets and Target Equipment.

b. Targets.—When regular printed targets are not available, suitable substitutes can be made on sheets of wrapping paper. Dimensions should be accurate. Improvised targets can be made in large numbers by improvising a stencil using heavy linoleum.

c. Ranges.—(1) *General.*—The range should be level and open and, if practicable, so located that fire can be delivered against a steep hill or bank in rear of the targets. Semipermanent bases should be constructed to facilitate placing targets in position and changing targets with the minimum delay or confusion.

(2) *Marksmanship course.*—Targets should be spaced from 3 to 5 yards apart. The depth of the range should be not less than 40 yards.

(3) *Field firing.*—Ground rising gently toward the targets with a hill in rear of the targets should be selected if available. Avoid shooting down hill. The range should be at least 50 yards wide by 100 yards long.

INDEX

INDEX

○